nal Malleus Mandible Maxillae
tal Sphenoid Stapes Temporal
Ethmoid Frontal Hyoid Incus L
n Ossicles Occipital Palatine Parietal Sp
nium Vomer Zygomatic • Cranium Ethmo
dible Maxillae Nasal Neurocranium Oss
pes Temporal Turbinate Viscerocranium
l Hyoid Incus Lacrimal Malleus Mandible
pital Palatine Parietal Sphenoid Stapes T
gomatic • Cranium Ethmoid Frontal Hyoi
asal Neurocranium Ossicles Occipital Pa
rbinate Viscerocranium Vomer Zygomati
nal Malleus Mandible Maxillae Nasal Ne
tal Sphenoid Stapes Temporal Turbinat
n Ethmoid Frontal Hyoid Incus Lacrimal
ium Ossicles Occipital Palatine Parieta
cranium Vomer Zygomatic • Cranium Et
Mandible Maxillae Nasal Neurocranium
d Stapes Temporal Turbinate Viscerocra
ontal Hyoid Incus Lacrimal Malleus Man
Occipital Palatine Parietal Sphenoid Sta
ner Zygomatic • Cranium Ethmoid Fronta
xillae Nasal Neurocranium Ossicles Occi
poral Turbinate Viscerocranium Vomer Z
ncus Lacrimal Malleus Mandible Maxilla
tine Parietal Sphenoid Stapes Temporal T
c • Cranium Ethmoid Frontal Hyoid Incus
Neurocranium Ossicles Occipital Palatin

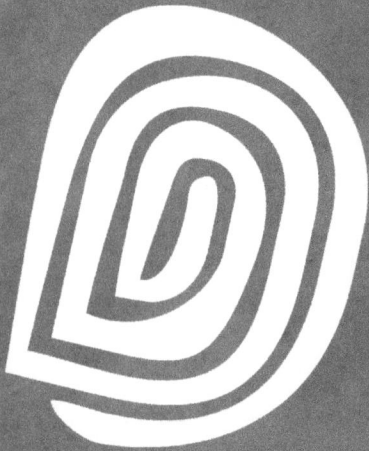

# Knowing Yourself Makes You Better at Everything

Dr. Bonyfide PRESENTS

For ages 6 to 206!

# BOOK 4

## BONES OF THE HEAD, FACE, AND NECK

A *KNOW YOURSELF, PBC* CREATION
OAKLAND, CALIFORNIA

Created by Know Yourself, PBC


www.knowyourself.com

ISBN 978-0-9912968-3-5
Library of Congress Control Number: 2015915711
Printed in USA

The skull is nature's sculpture.

David Bailey

Acknowledgements

Thanks to the many people without whose help this project would not have been possible. Their experience, educational and editorial expertise, network of resources, and unflagging enthusiasm helped make this book what it is. To our team at Know Yourself and our fantastic network of friends and advisors: Thank You!

# Contents

# Preface

Know Yourself, PBC, is a Self Literacy company. We are on a mission to prepare all young people to understand their self worth.

We believe that the path to understanding self worth is Self Literacy, which we define as a foundational knowledge of anatomy, physiology, and psychology, as well as how these three things work together.

We know that self worth grounds human beings in themselves. Being grounded allows for self advocacy and having an awareness of who you are in your community and on this earth.

To this end, we create products that guide young people on the path to being Self Literate. They include award winning books, activity kits, toys, and comics, which are all super-fun and accessible educational materials.

# Welcome!

Want to know your body's secrets?
It's amazing what's inside!
I'll help you learn about yourself.
Hello, I'm Dr. Bonyfide!

Come on a guided tour with me.
Know yourself inside and out.
We'll start with the framework of bones,
and you'll learn what they're all about.

Your skeleton is a puzzle.
And these parts, they interlock.
But they're not puzzle pieces;
each bone is a building block.

Yes, these bones have names and functions
for you to learn about.
You may think it's overwhelming
But my fun rhymes will help you out.

And before you even know it,
you'll have learned them all with ease.
You'll sing and say the names of bones
just like your ABCs!

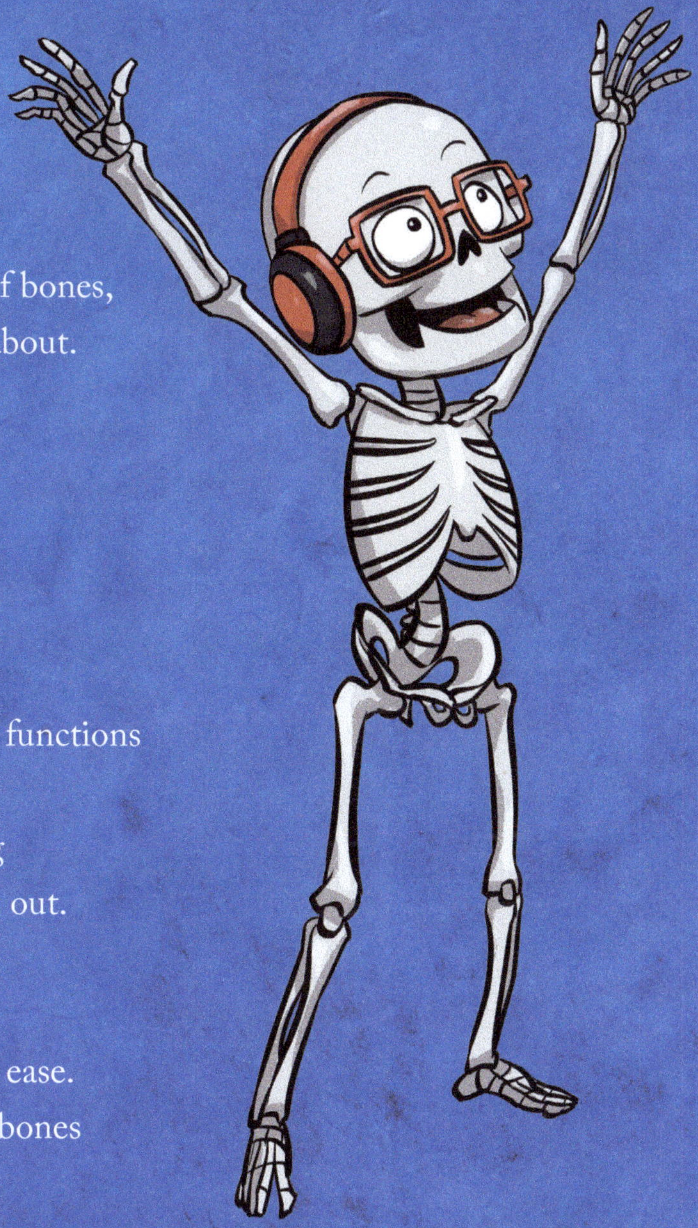

# The Skeleton

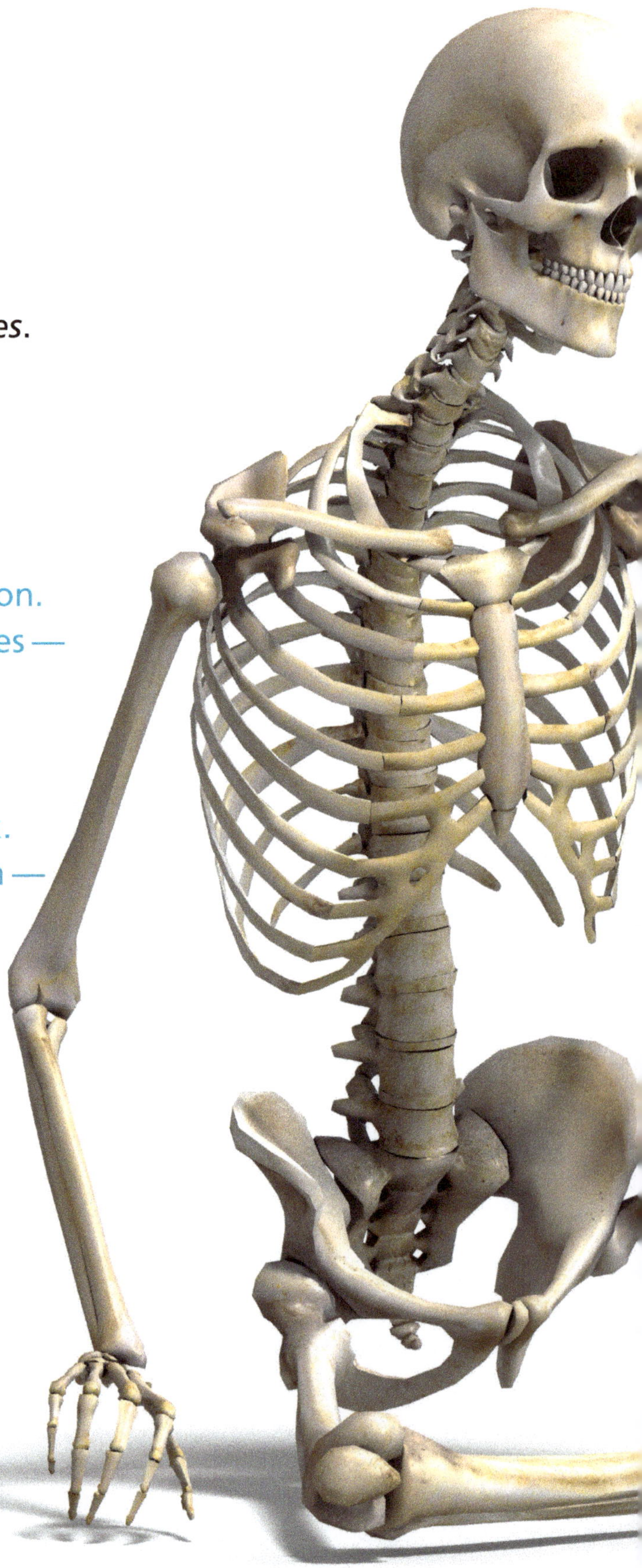

**Osteology** means the *study of bones*.

You see, *-ology* at the end of a word means "the study of."

When you grow up, as you'll find out, you'll have **206** bones in your skeleton. Think how you'd be without your bones — squishy, just like gelatin.

Book 4 explores the **29** bones that are found in your head, face, and neck. We're going to learn about all of them — it will be an exciting trek!

Say them like this:
Osteology "**os-tee-OL-uh-jee**"
Osteologists "**os-tee-OL-uh-jists**"

The stressed syllable is always shown in **CAPITALS** and **red**.

## How to use our Pronunciation Guide

To help you learn how to pronounce new words, we've invented a fun system. We separate words into syllables, which is something you've probably seen before, but whenever a syllable sounds like a word you might be familiar with, we spell out the actual word. For example, the word palatine would be shown as

"**PAL-uh-tine**"

This pronunciation includes the words "pal" and "tine."

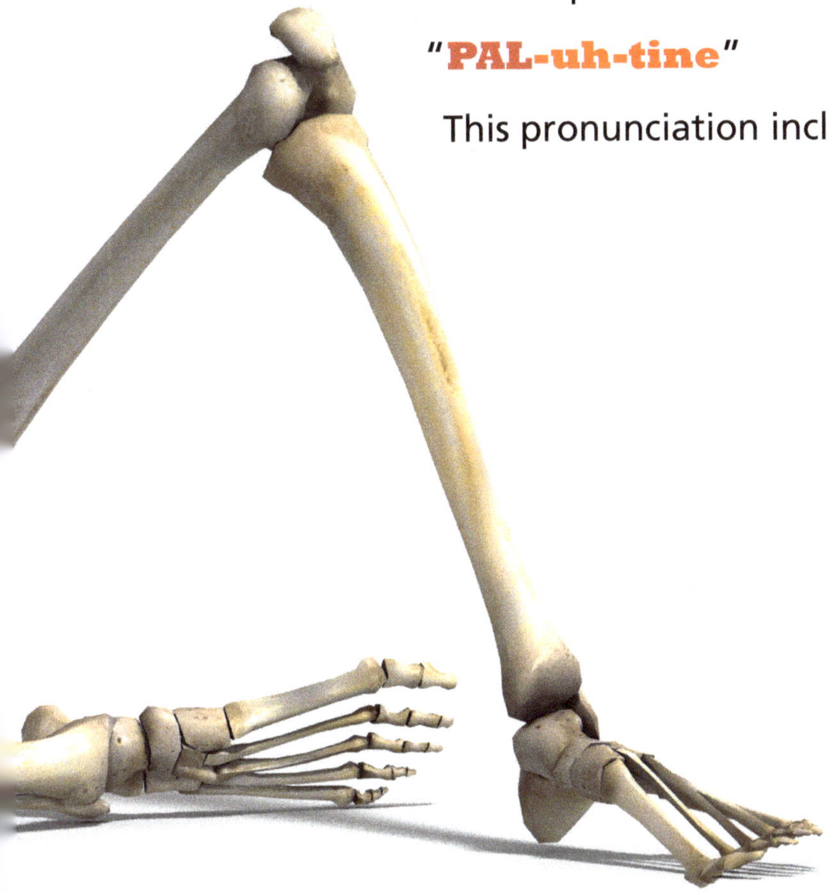

## Let's start with the big picture.

Touch your head,
now your chin.
There's something hard
underneath your skin.

What is it exactly?
You are feeling your bones—
your living skeletal framework.

Humans are born with approximately **300** bones. As we grow, some bones fuse together, leaving us with about **206** bones in the adult human body.

What do you think your bones do?

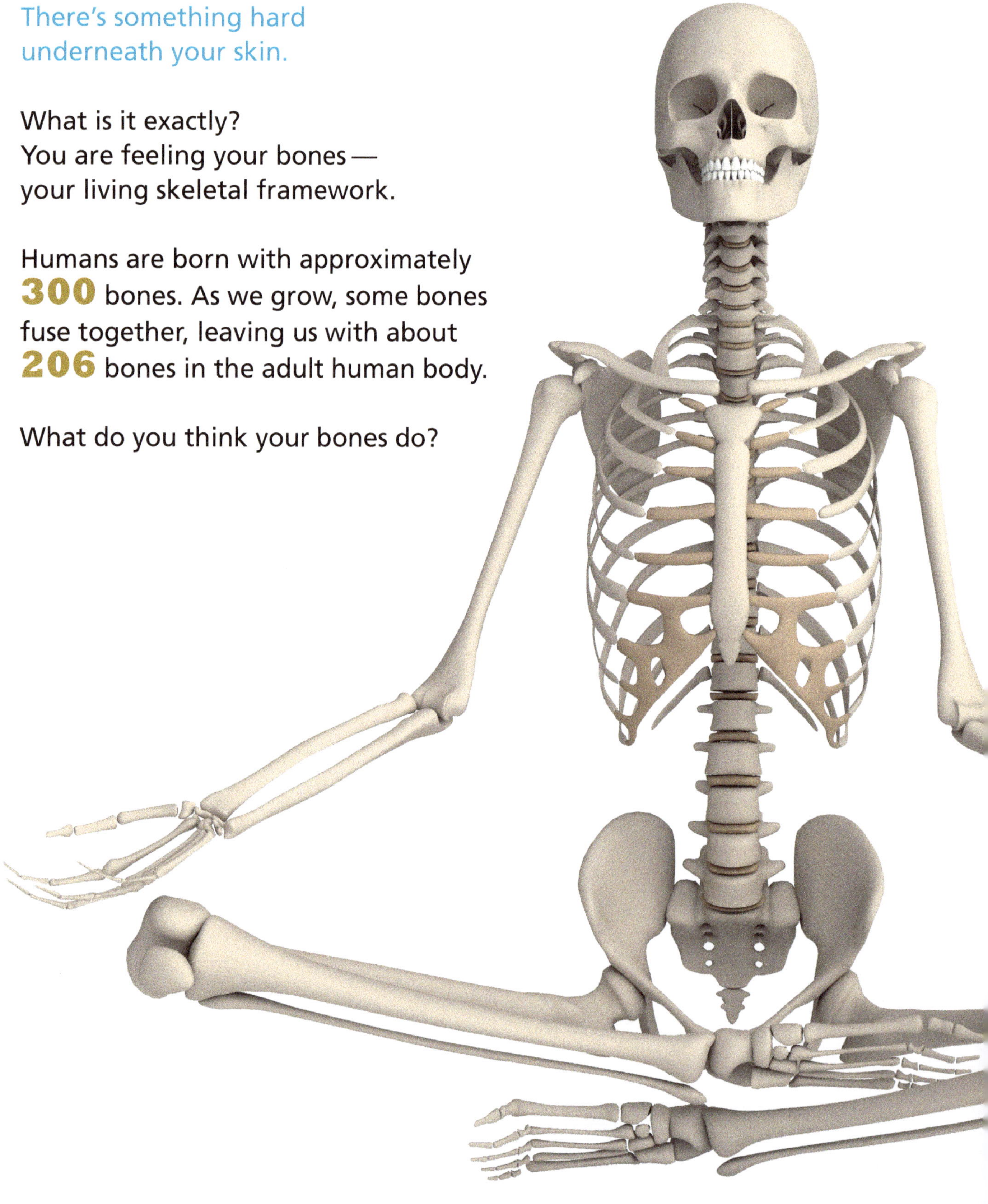

The bones of the skeleton have five main functions:

1. **Structure**: The organization of the bones in the skeleton give your body its shape.

2. **Protection**: Bones protect several organs, such as the heart, lungs, stomach, and intestines. The bones of the head, face, and neck protect the brain and the spinal cord.

3. **Movement**: Muscles are attached to bones. Together, muscles and bones make it possible to move your body.

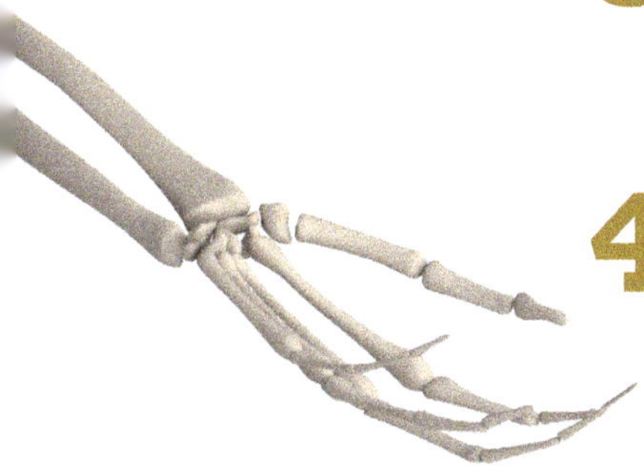

4. **Production**: Red blood cells, which transport materials, and white blood cells, which protect against disease, are made in some bones of the body.

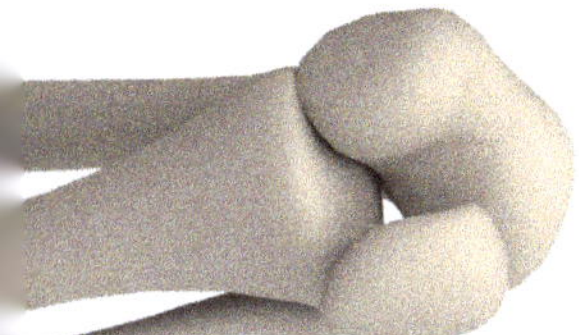

5. **Storage**: Minerals such as calcium and phosphorous are stored in bones.

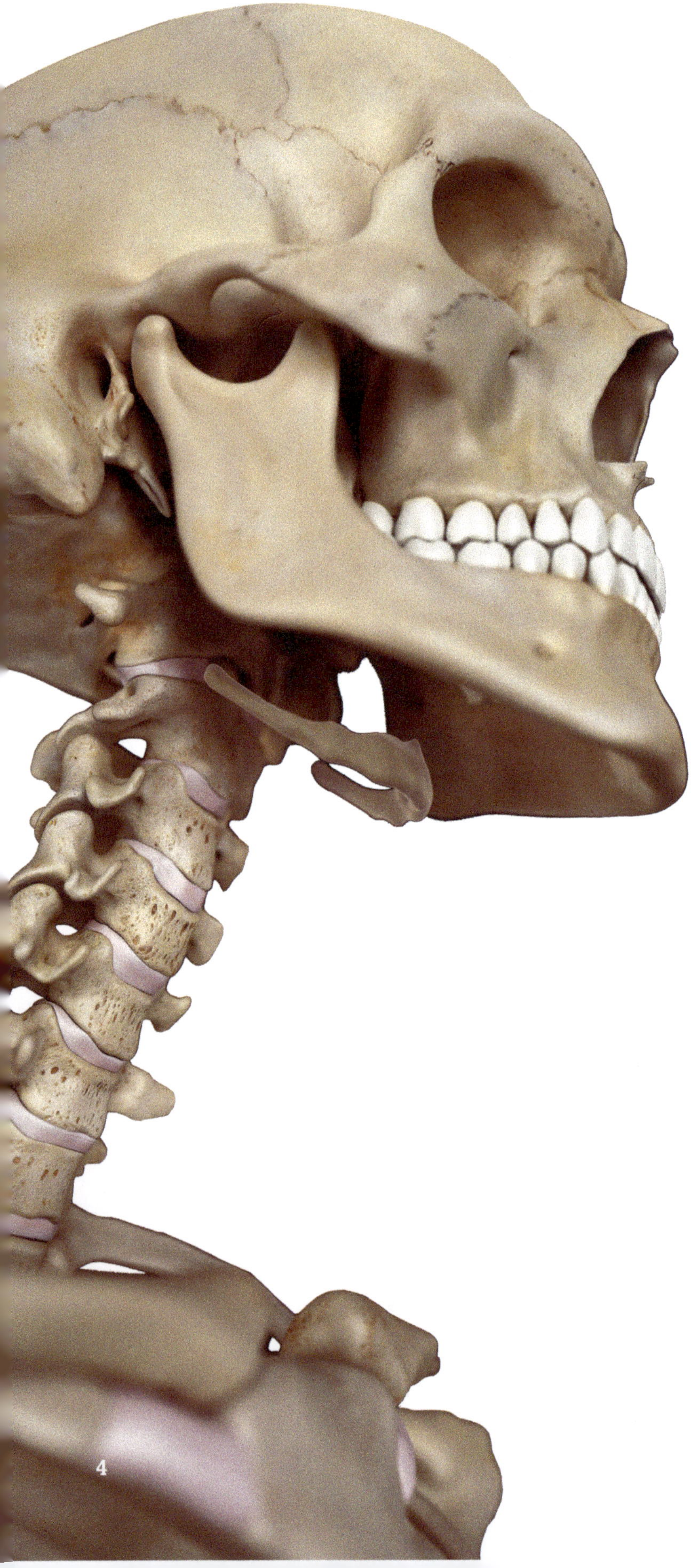

Bonyfide Buddies,
we've got a fun-tastic trip in place.
Get ready to explore
the bones of your head, neck,
and face.

Put your hands on your head.
Now feel your whole face.
Behind the skin sits your skull,
your brain's protective case.

While the back of your skull
forms a large dome,
the front is where your eyes
and your nose make their home.

Believe it or not,
your ears have some bones.
They help you make sense
of sounds and of tones.

There's also a bone
hanging out in your neck.
It supports your tongue's weight
and keeps your voice in check.

# Head, Face, and Neck

Your skull—and the bones in and around it—are really cool.

- Your skull houses your brain. An adult brain weighs about 1.3 kilograms (3 pounds).
- The average adult head has a mass of 4.5 kilograms (about 10 pounds), including the brain, muscles, bone, teeth, eyes, and other organs and tissues.
- The bones of your skull average about 0.63 centimeters (0.25 inches) in thickness.

That's a lot of weight sitting on top of your neck!

The joints between the bones of the skull are called sutures. Sutures are immobile joints, which means they don't move. They are found only in the skull. When you were born, the sutures in your skull were wider because the bones weren't fully formed. The sutures are filled with connective tissue, leaving areas that are called "soft spots." Over the first 18 months or so of your life, the connective tissue is replaced by bone, and the soft spots disappear.

In this book, you're going to learn the names and locations of the 29 bones of your **head**, **face**, and **neck**.

Count them up like this!

| | | |
|---|---|---|
| | 14 | bones in the face |
| | 8 | bones in the head |
| | 6 | bones in both ears |
| + | 1 | bone in the neck |
| | ______ | bones of the head, face, and neck |

Okay, let's go explore...

The real voyage
consists not
but in

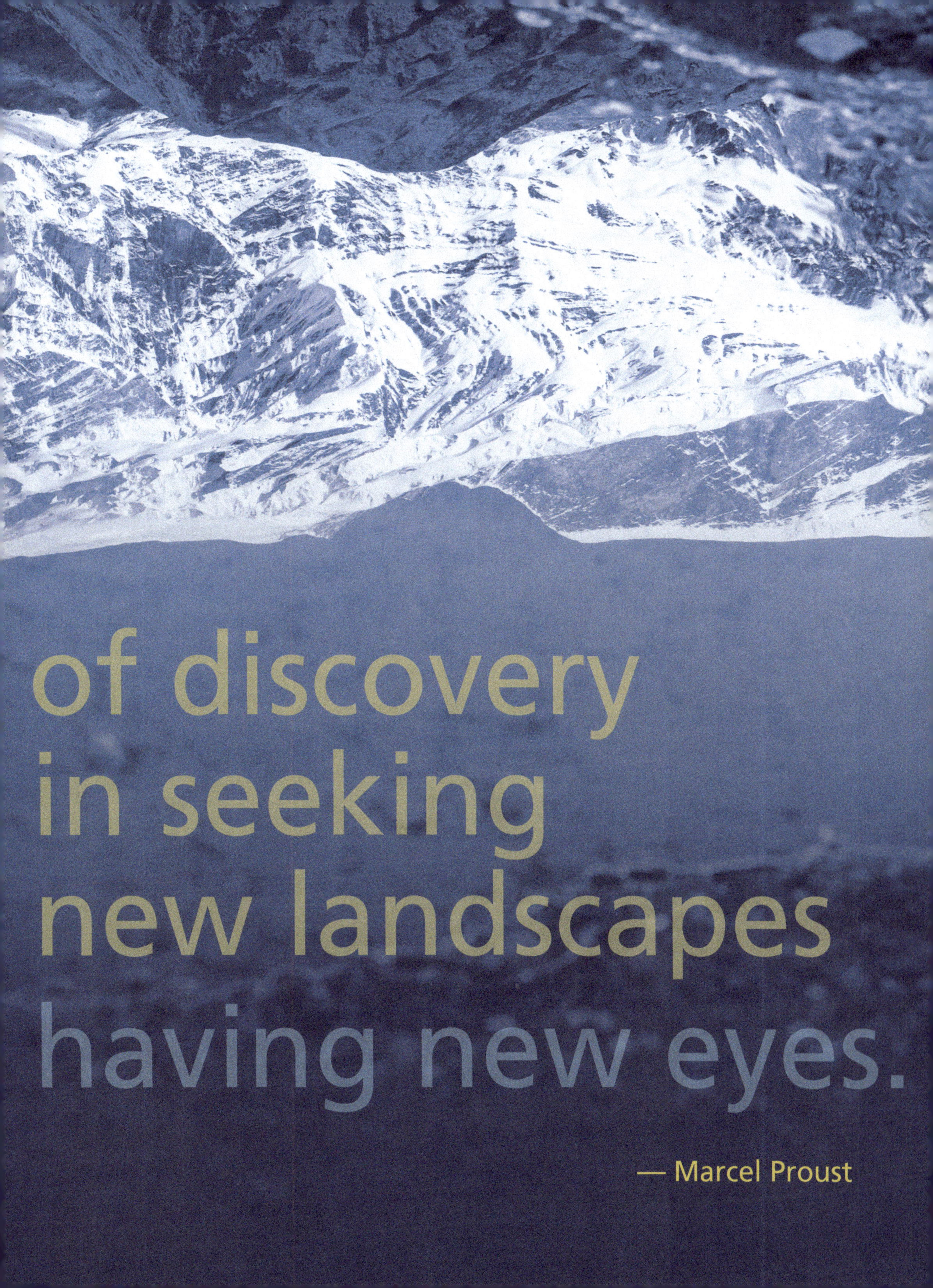
of discovery
in seeking
new landscapes
having new eyes.
— Marcel Proust

Lake Merritt
Oakland, California
Caw?
First, shape your dough into a skull.
Next, use your fingers to shape the eye sockets, nose, and mouth.
You're done!
Hey, Pinky! What are you making?

I have to make this calavera for my friend Stokely's party tonight.
I just can't get it to look right, and I'm running out of time!
To get your skull looking right...
...you have to know it from the inside out...
...this sounds like a job for...
Captain B!
Oh boy...
That's not a boy, it's my trusty sidekick, Seymour Seagull.

Welcome to Skullapagos Island!
That sounds scary!
Don't worry, we have a map to find our way around!

MAXILLA GORILLA
SEYMOUR SEAGULL
OLIVER OSSICLE
N
E
W
S
LADY LANA LACRIMA
OCCIPITAL OX

# Overview of the

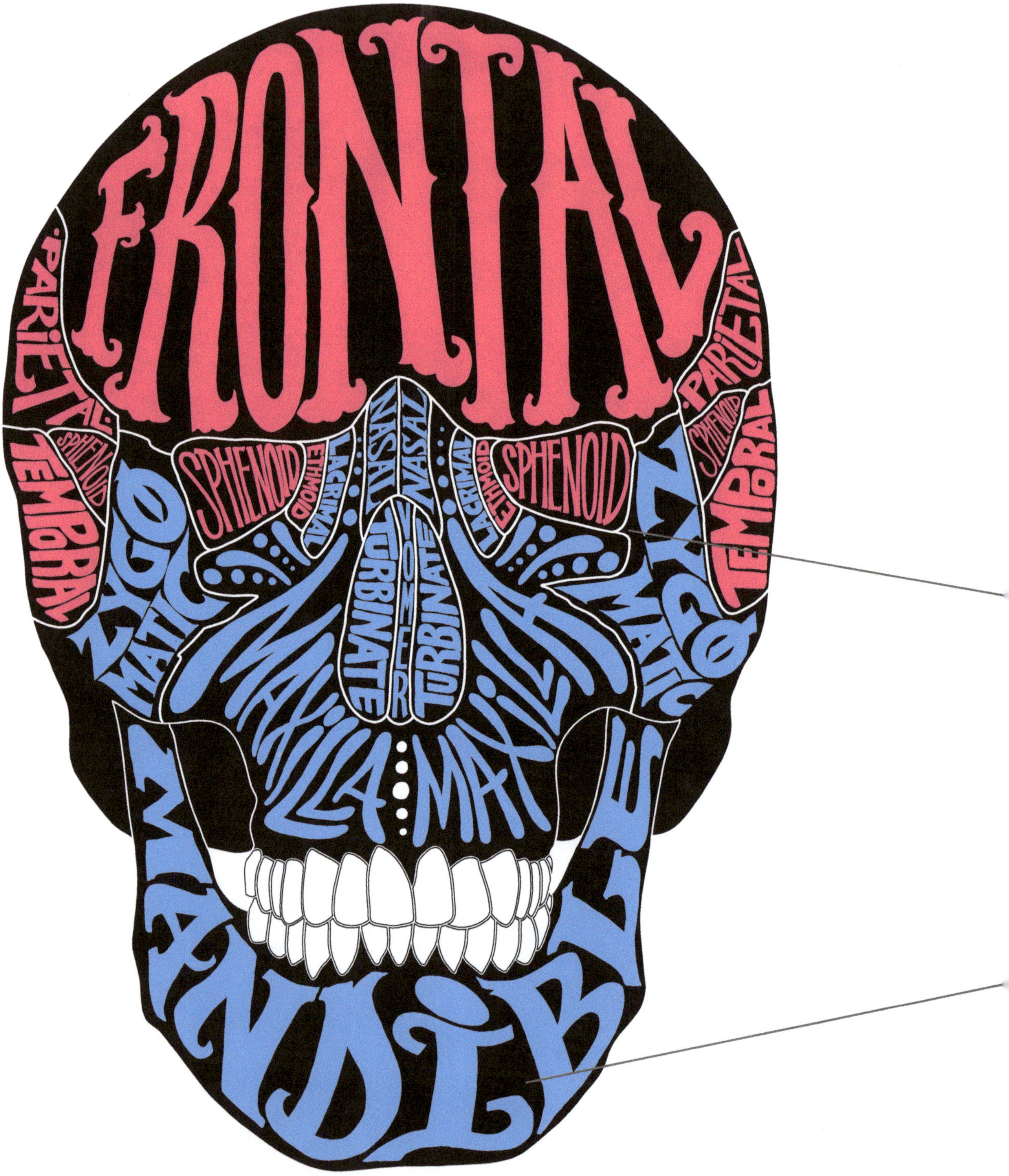

**Heads Up** *This illustration does not include the palatine bones because they form the back of the roof of the mouth, inside the head.*

# Head, Face, & Neck

Our skulls are classified into two main sections:
the **viscerocranium** (face), which has **14** bones, and
the **neurocranium** (head), which has **8** bones.

Say them like this:

"**viss-uhr-oh-KRAY-knee-uhm**"
"**nuhr-oh-KRAY-knee-uhm**"

In all, your head, face, and neck contain 29 bones. These bones protect the brain and many of the organs that allow us to taste, hear, smell, see, and touch.

The skull also contains the **ossicles**, or bones of the ear (**3** bones in each ear), and the **hyoid**:

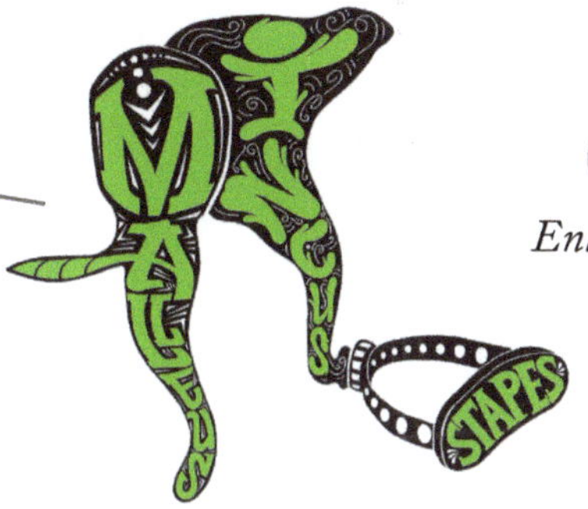


Ossicles
*Enlarged View*

Hyoid

Your head is ROUND like a globe, so we need to see it from the side . . .

*Side View*

. . . and from the back. This way we get a better look at all the bones of the skull!

*Back View*

*Enlarged View*

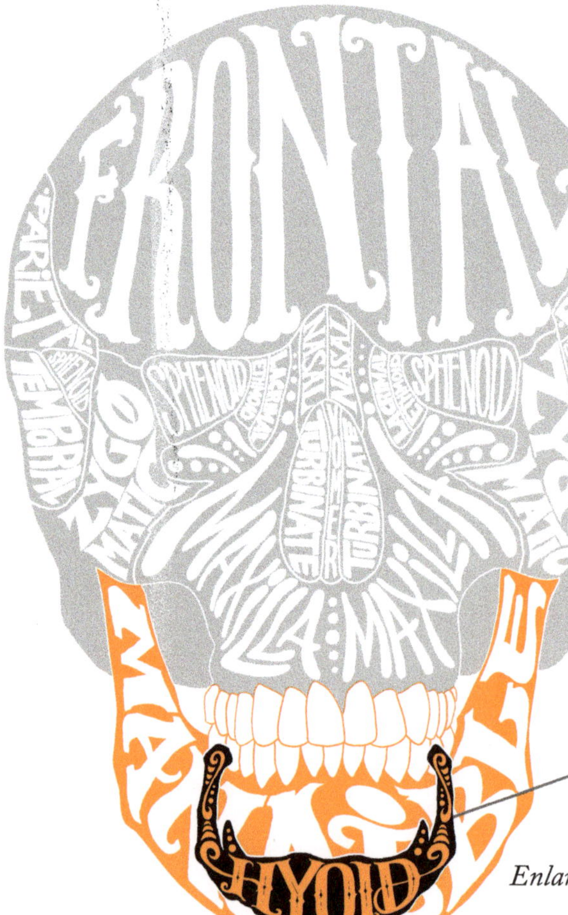


*Enlarged View*

# These bones are teeny weeny.

The **ossicles** are the bones found in the ears. Each ear has three ossicles. They are considered part of the skull, even though they are in neither the viscerocranium nor the neurocranium.

Say it like this: "**AWE-sick-uhl**"

The average earlobe is 2 centimeters, or 20 millimeters (0.8 inches) across, and the average stapes is 3 millimeters (0.1 inches).

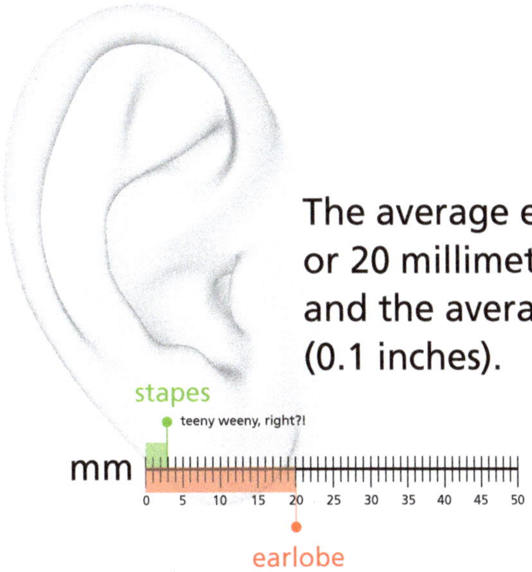


The **hyoid** bone is also considered part of the skull, even though it's found in the neck. It lies behind the mandible and plays a role in movement of the tongue and swallowing.

Say it like this: "**HIGH-oyd**"

**Heads Up** *The smallest bone in the human body is the stapes.*

# Paired Bones? Two of a Kind!

Like socks on your feet or gloves in cold weather,
they're a natural match, you will find.
Like bookends or best friends, they're better in twos,
and paired bones are two of a kind!

Throw a party with your **parietal** partners,
a wall-to-wall jubilee!
Sound the drums for your **temporal** team.
Time for cranial festivity!

Face up to your **conchae** companions
which play a role in air blown.
Don't cry—your **lacrimal** duet
won't let you shed tears alone.

Say hello to your **maxillae** amigos,
and go grab a bite to eat.
Then there's our distinctive **nasal** alliance—
two neighbors that smell so sweet.

Should we go visit a chateau?
Just **palatines** passing the time...
Eye to eye and cheek to cheek,
**zygomatic** mates forever...two of a kind!

# Activity: Skull Symmetry

Help Pinky fix the skull by using the grid to finish the half skull.
See how accurately you can finish the rest!

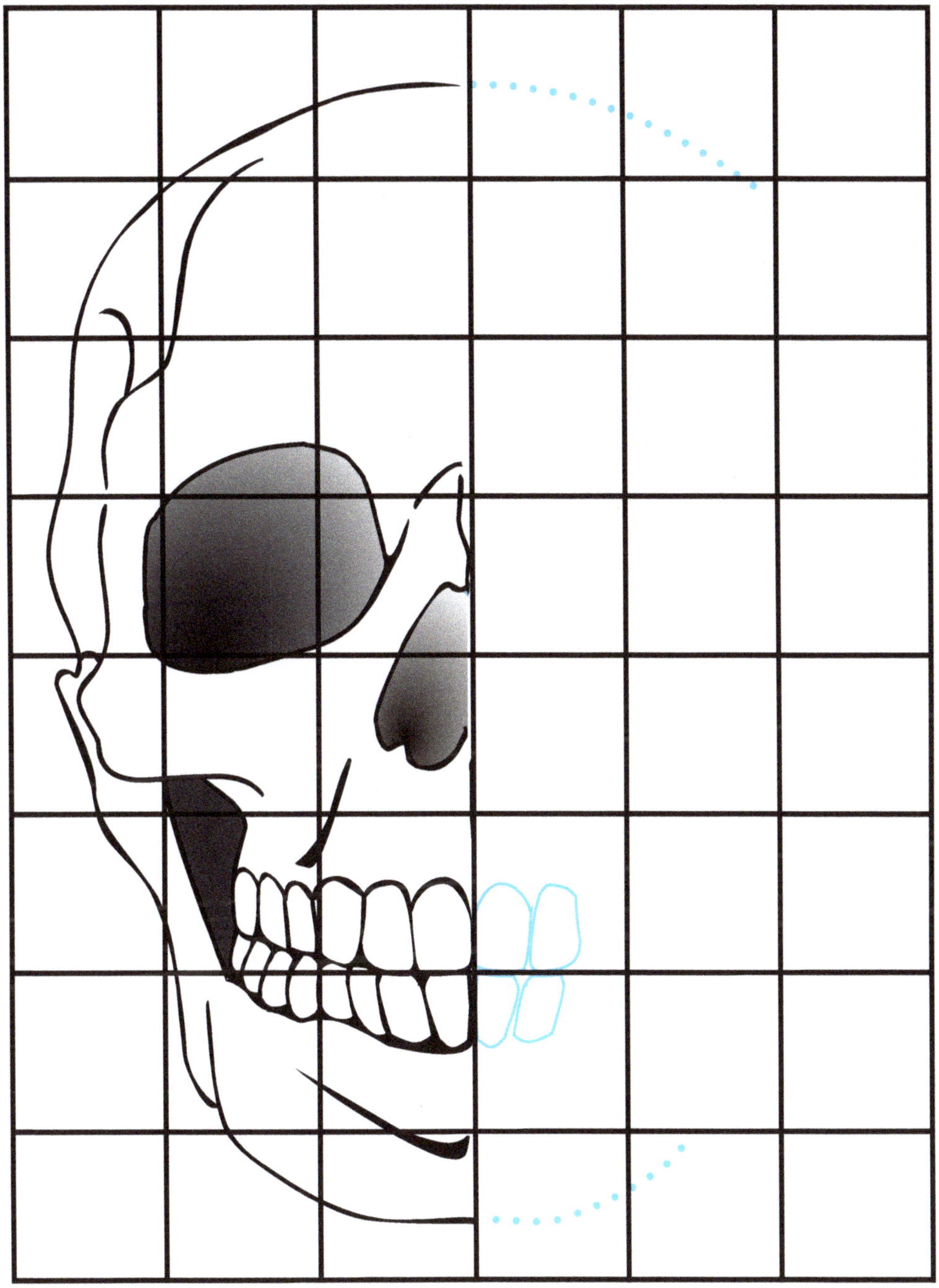

# Body Symmetry

Humans—and Seymour Seagull—have bilateral body symmetry. That means we have two sides, or halves, that are mirror images. If you draw a line down the middle of the front of the human body from top of the skull to feet, the bones on each side of the body, including the skull, are similar in size, shape, and position.

Say it like this: "**bye-LAT-uhr-uhl SIM-uh-tree**"

**Heads Up** *What is not symmetrical about Seymour Seagull?*

# Overview of the

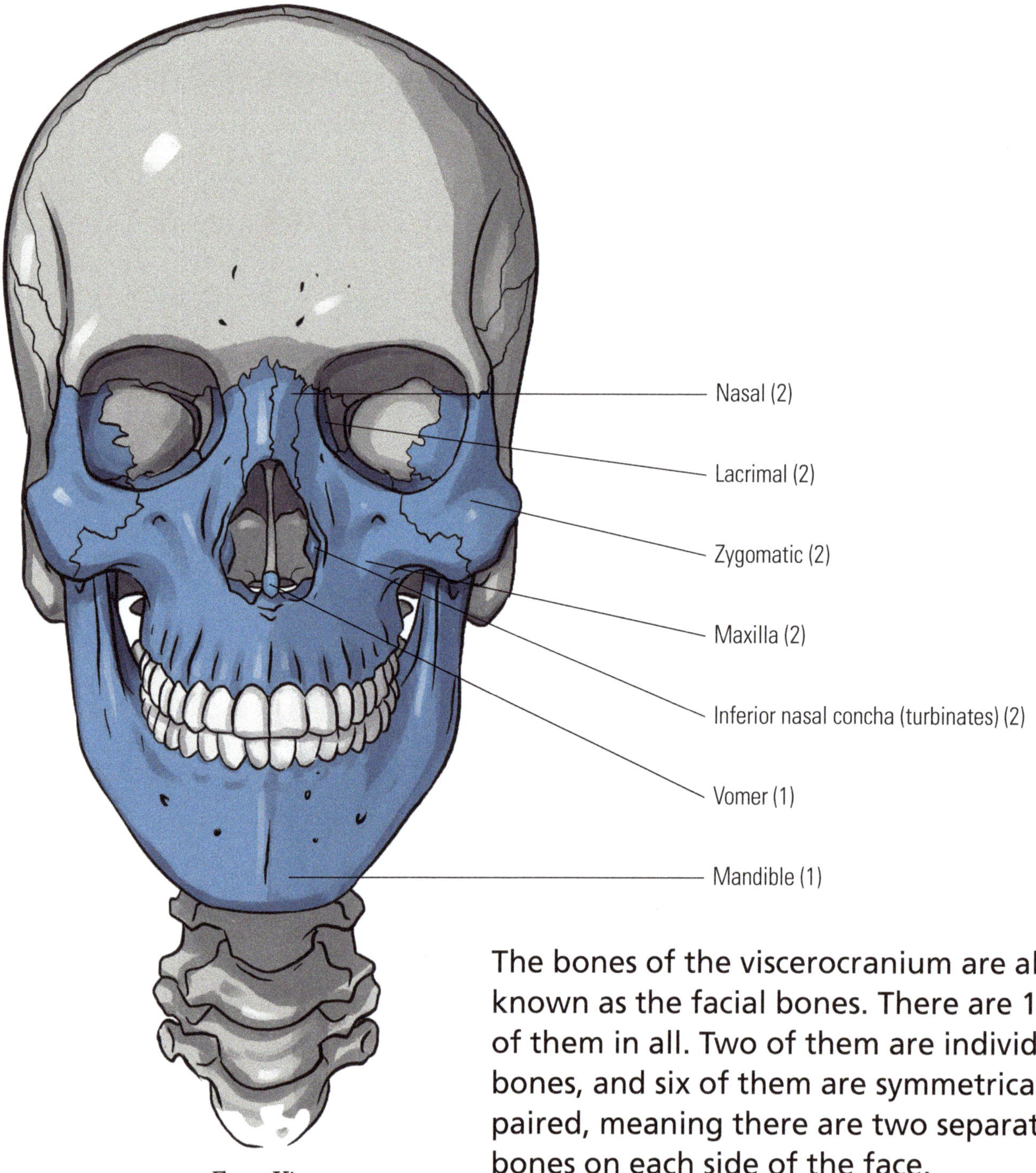


*Front View*

The bones of the viscerocranium are also known as the facial bones. There are 14 of them in all. Two of them are individual bones, and six of them are symmetrically paired, meaning there are two separate bones on each side of the face.

**Heads Up** *"Cranium" comes from the medieval Latin word for skull. The prefix "viscero" is from the Latin* viscera, *which refers to the internal organs. The viscerocranium contains organs such as your tongue and tonsils and forms part of your eye sockets.*

# Viscerocranium

You may have noticed a pair of bones missing in the two images shown here. The bones are called palatine bones, and they help form the back of the roof of the mouth and other spaces in the skull. The vomer and inferior nasal conchae (turbinates) aren't visible from the side.

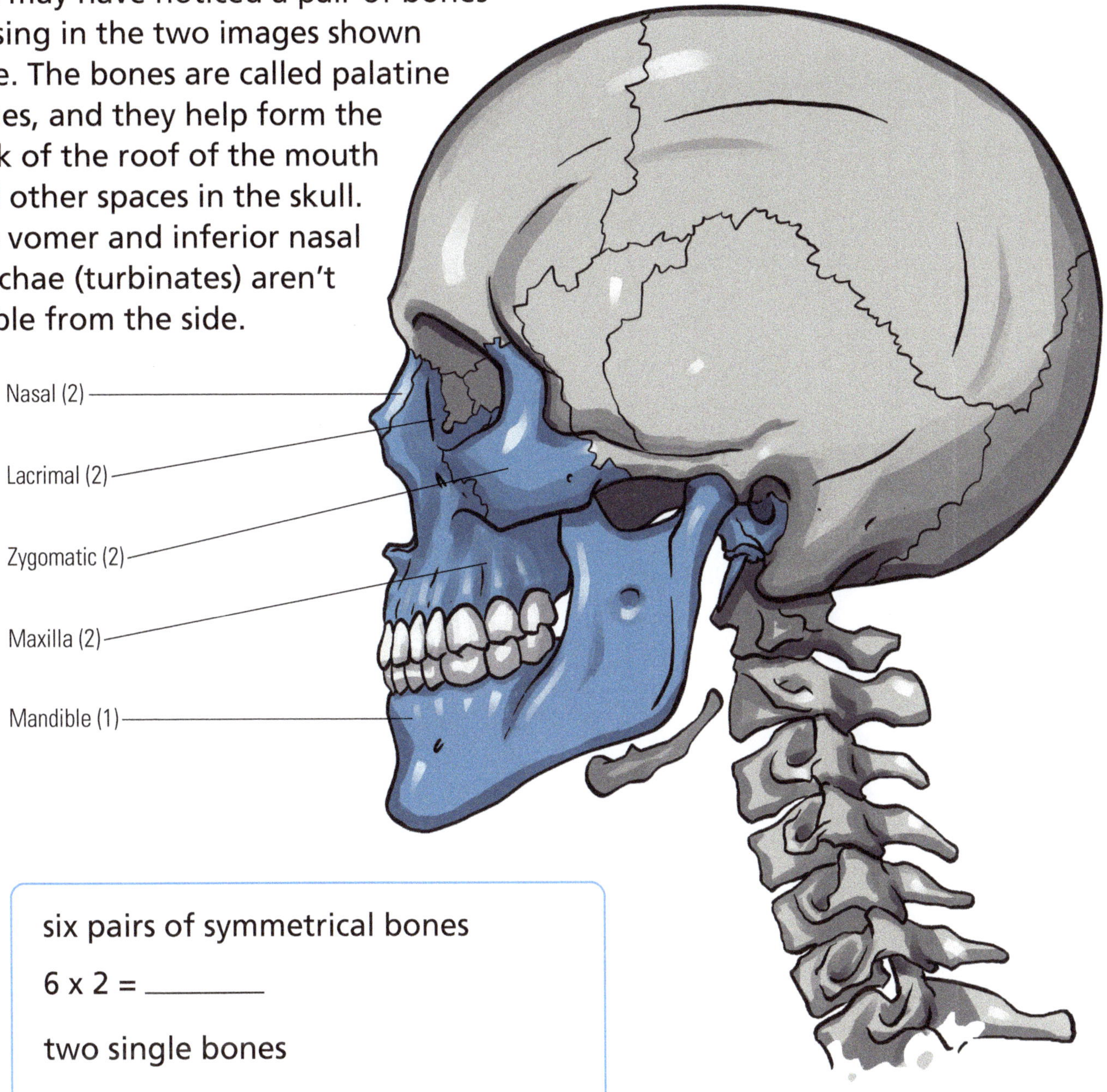


*Side View*

six pairs of symmetrical bones

6 x 2 = ________

two single bones

1 + 1 = ________

________ paired bones

+ ________ single bones

________ viscerocranium bones

# Mighty Pinky Moves Vertically

**Heads Up** *Part one of the viscerocranium section utilizes the first four words of our mnemonic: M, P, M, and V.*

# ZIP LINING NEAR TREASURE

You're about to learn the names and functions of all the bones in the viscerocranium. Here's a mnemonic (word trick) that will help you remember their names.

Say it like this: "**ne-MON-ick**" (the first "m" is silent)

ZIP — Zygomatic

LINING — Lacrimal

NEAR — Nasal

TREASURE — Turbinate

First stop, the mandible!
In there?

# Mandible

Say it like this: "**MAN-duh-bull**"

The **mandible** is also known as the jawbone. It's the strongest, largest, and lowermost bone in the face. It is also the most moveable bone in the skull, mostly moving up and down, but it can also move side to side.

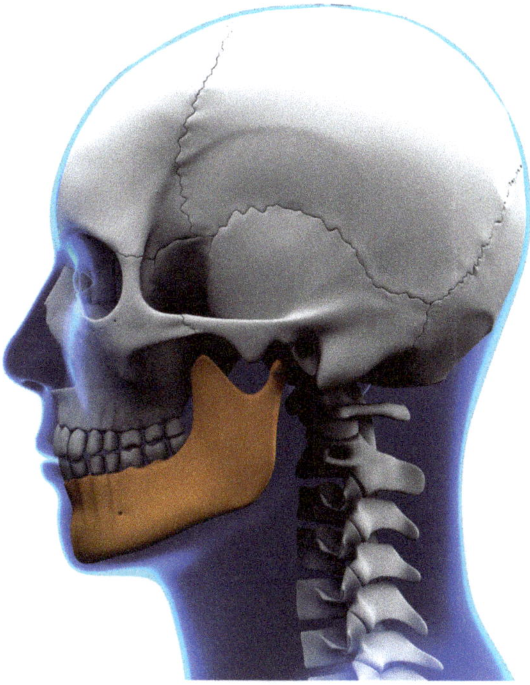

*Side View*

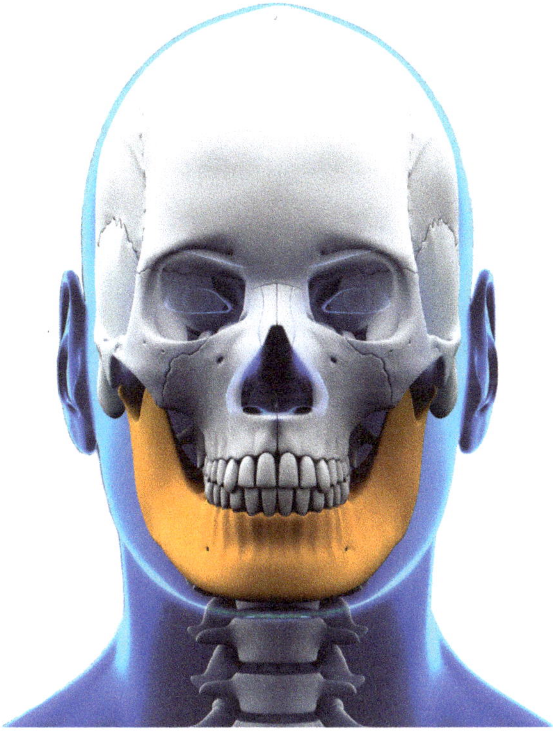

*Front View*

# Etymology

Say it like this: "**et-uh-MOL-uh-jee**"

Etymology is the study of the origin of words.

Mandible comes from the Latin word *mandere*, which means "to chew."

# Bone Info

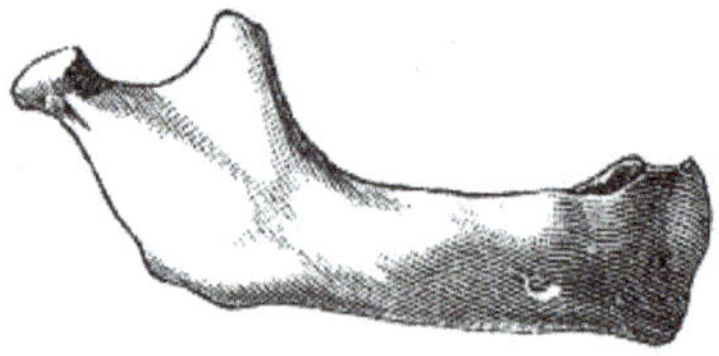

*Infant*

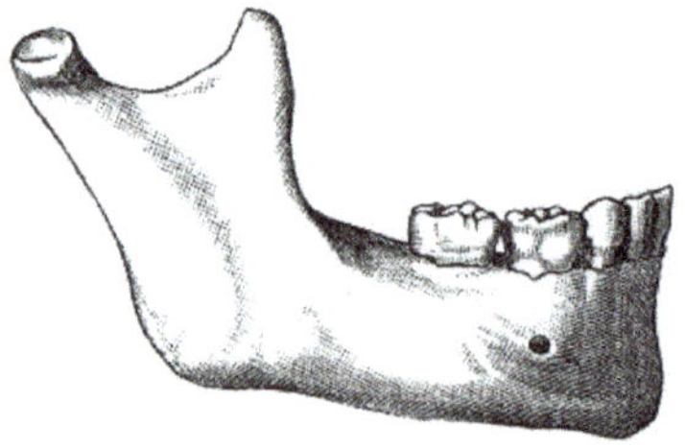

*Child*

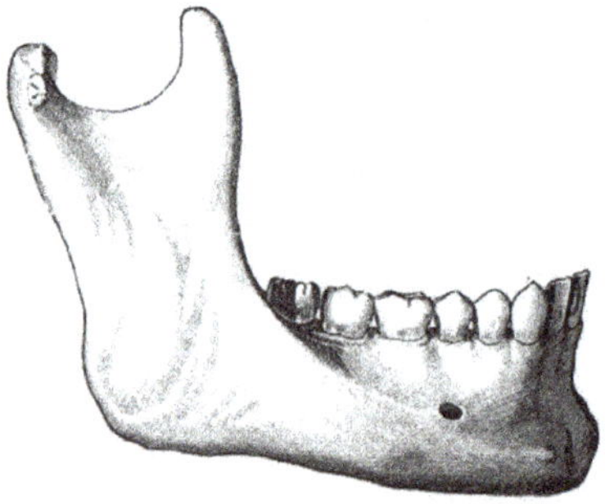

*Adult*

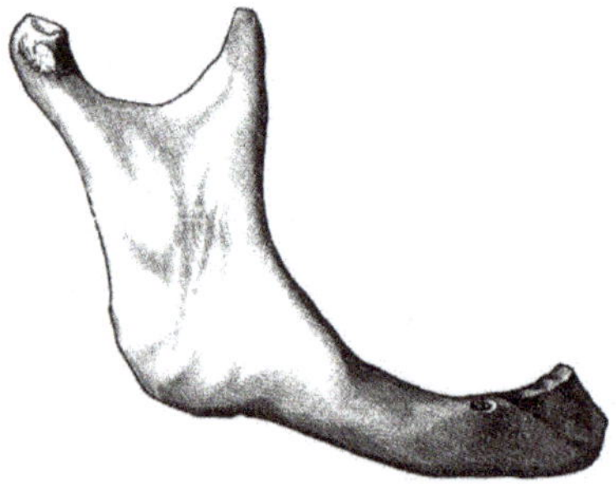

*Elderly*

In addition to biting, your mandible (also known as the lower jaw) helps you chew and talk. It holds your bottom teeth in place, too.

The mandible is also a time teller! Like a few other bones in the body, it doesn't fully form until adulthood, and then it begins to decay as you grow old.

An archaeologist, or a scientist who studies human history by looking at artifacts and remains, can use the mandible to estimate the age of an individual when he or she died.

# Activity: Chew on This! Part I

When you eat, you use your **temporomandibular joints**, (shortened to **TMJ**).

Say it like this: "**TEM-pore-oh-man-DIB-you-luhr**"

The TMJ joins your temporal and mandible bones, which means you have two of these joints, one on each side of your face. They are among the most frequently used joints in your body.

(The temporal bones are coming up soon in your adventure!)

Let's find your TMJs!

1. Put your fingers just in front of the middle of your ear at the side of your head.
2. Now, open and close your mouth.
3. Do you feel a bony hinge moving underneath your skin? Those are your TMJs!

Time to explore the palatine, know what I mean?

# Palatine

Say it like this: "**PAL-uh-tine**"

The **palatine** is a paired bone located toward the back of the mouth. It helps to form the roof of the mouth and the nasal cavity, a large air-filled space above and behind the nose in the middle of the face. It also forms part of the orbital floor, or eye socket.

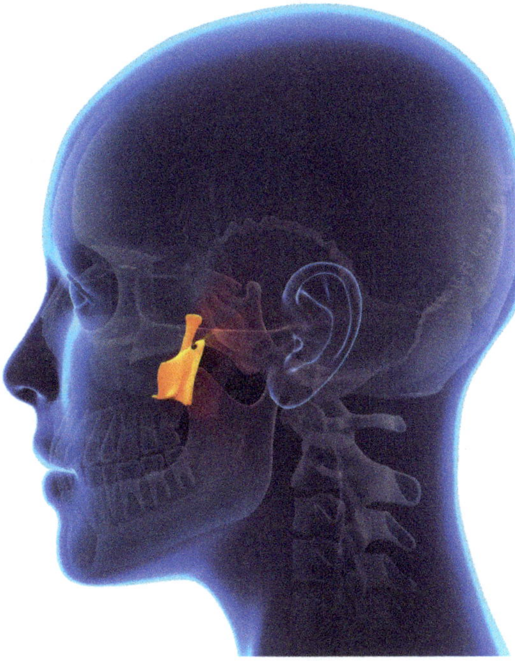

*Side View*

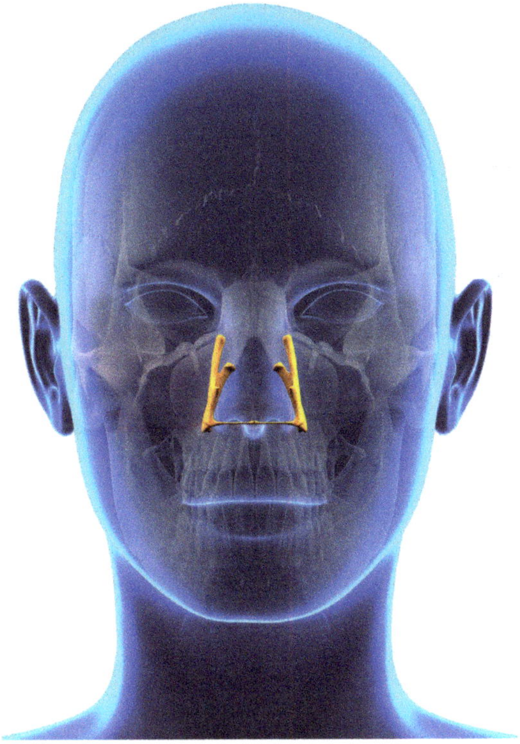

*Front View*

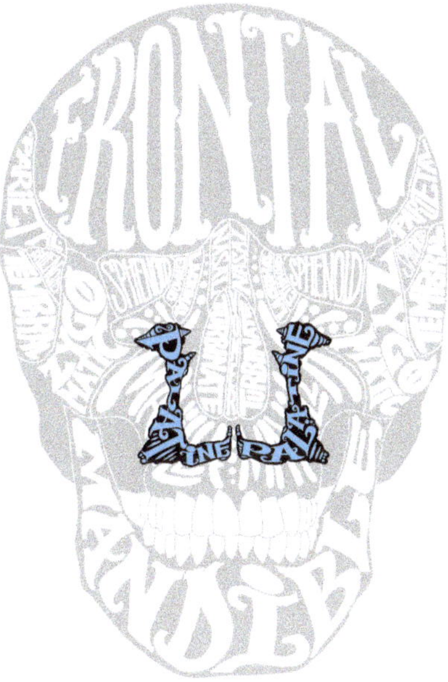

**Heads Up** *Remember, when we say "paired bone," we really mean two bones that are symmetrical.*

Mighty Pinky Moves Vertically Zip Lining Near Treasure

# Etymology

Palatine comes from the Latin word *palatinus*, meaning "of the palace."

The palatine bone forms
the "roof" of the mouth.
PINKY'S
PALACE


# Bone Info

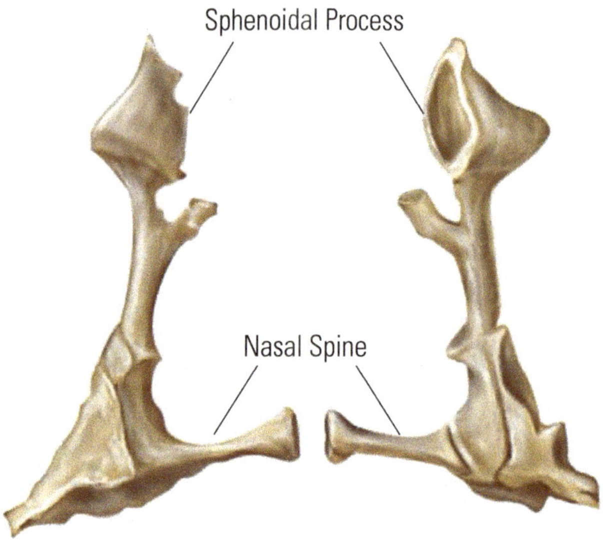


*Front View* *Back View*

With the maxilla bones, the palatine forms the hard palate.

Make a clicking noise by placing your tongue against the top of your mouth and snapping it back quickly. Try it a few a times.

Click. Clack. Clack.

What you feel is your palate, otherwise known as the roof of your mouth. It has two parts, the hard part in front near your teeth and lips, and the soft part in back near the tonsils.

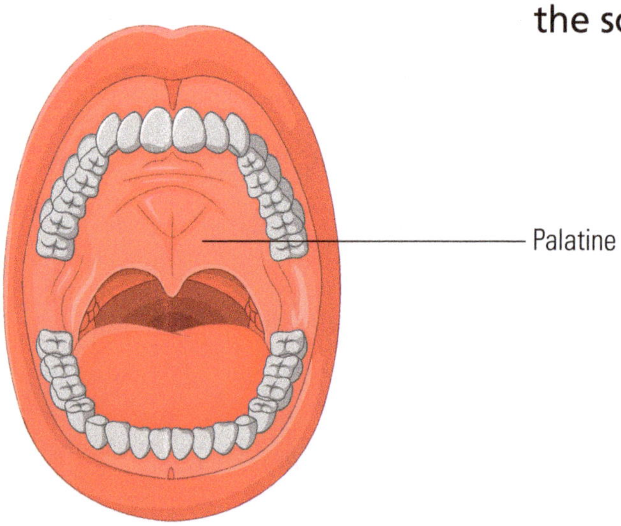

# Activity: Reservations for Two at Pinky's Palace?



Nestled in the mandible is the tongue. The human tongue has four types of **papillae** which look like small bumps. Three of them contain clusters of taste buds.

Say it like this: "**puh-PILL-ee**"

Taste buds have sensitive microscopic hair-like structures called **microvilli**. These microvilli send messages to the brain, which the brain interprets as taste.

Say it like this: "**my-kro-VILL-eye**"

What are three of your favorite flavors, and what foods do you find them in?

| Tastes | Foods |
|---|---|
| | |
| | |
| | |

What was that? It sounds like it came from upstairs.
That's got to be Maxilla Gorilla! Let's go find her.

# Maxilla

Say it like this: "**mack-SILL-uh**"

The **maxilla** is a paired bone that forms the upper jaw and part of the palate.

When referred to in the plural, the bones are called the maxillae.

Say it like this: "**mack-SILL-ee**"

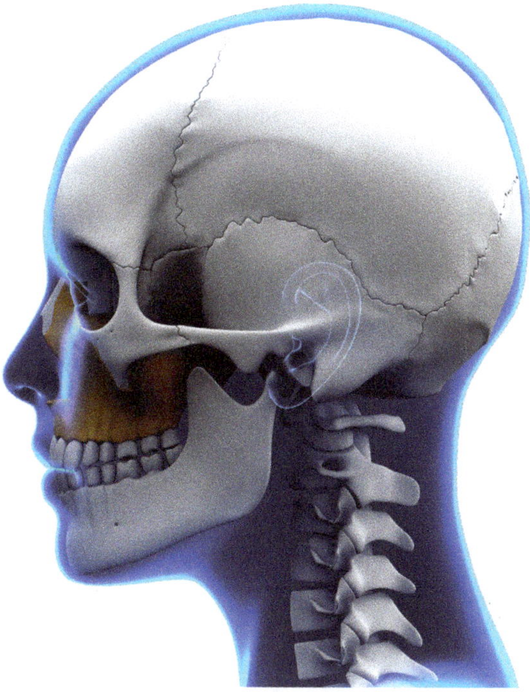

*Side View*

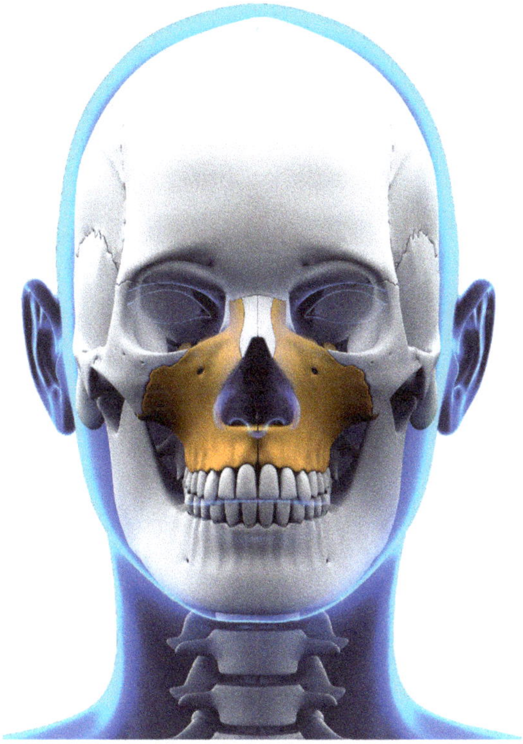

*Front View*

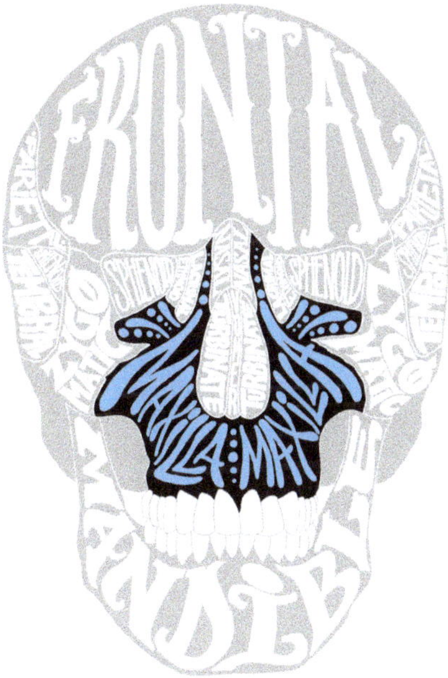

# Etymology

Maxilla comes from the Latin word for "jaw." The word is related to the Latin word *malaris*, which means 'cheekbone.' The upper part of the maxilla forms a small part of your cheek, right under your eye.

# Bone Info

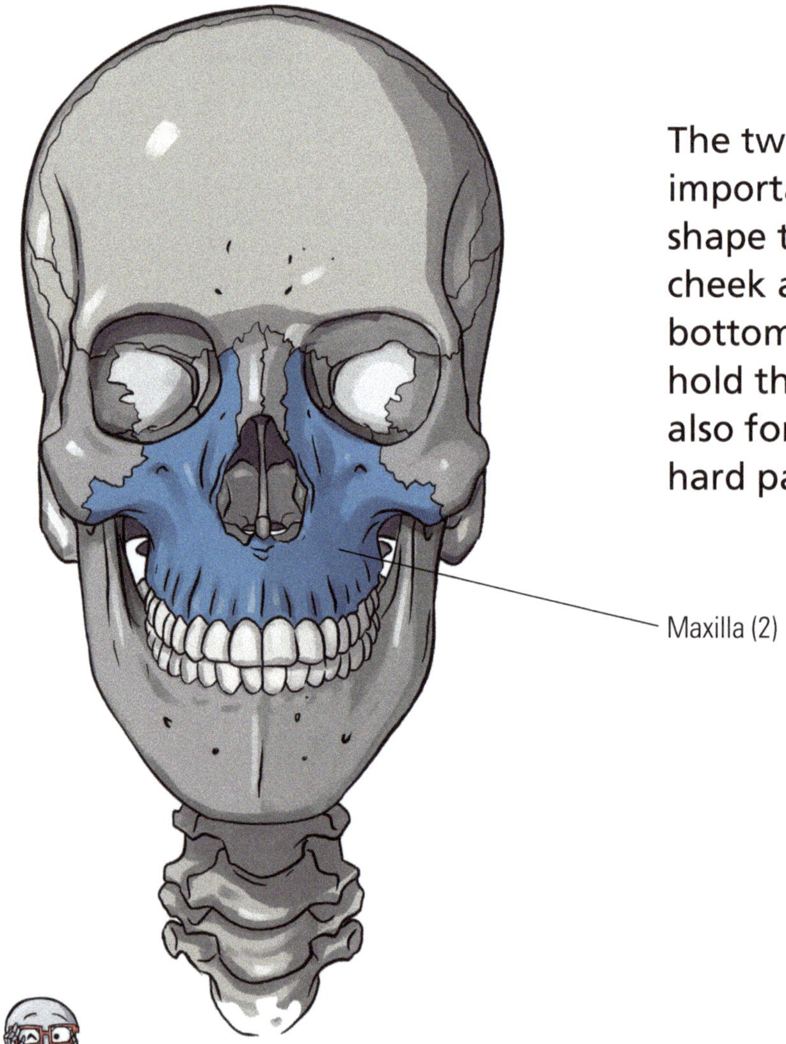


The two maxilla bones have important functions. They help shape the face, along with the cheek and nose bone; form the bottom of the eye socket; and hold the top teeth. The maxillae also form the front part of the hard palate.

**Heads Up** *Some of the text so far references bones that are on your journey ahead!*

# Activity: Agree or Disagree?

## Seymour Seagull Wants to Caw-Know…

| | Agree | Disagree |
|---|---|---|
| Seymour Seagull is Dr. Bonyfide's trusty sidekick. | ☑ | ☐ |
| The bones of the skull are classified in two main sections. | ☐ | ☐ |
| 14 + 8 + 6 + 1 = 30 | ☐ | ☐ |
| The mandible is a single bone in the viscerocranium. | ☐ | ☐ |
| A mnemonic is a memory tool that can be used to remember the names of the bones of the head, face, and neck. | ☐ | ☐ |
| The palatine is a famous palace. | ☐ | ☐ |
| People have bilateral symmetry, which means many bones of the skull are paired. | ☐ | ☐ |
| The Latin root for mandible refers to internal organs. | ☐ | ☐ |
| The maxilla is a paired bone that makes up the lower jaw. | ☐ | ☐ |

Mighty Pinky Moves Vertically Zip Lining Near Treasure

Dr. B, where are we now?
Maxilla! Why aren't you tending the lettuce beds?
Come on in, and let me tell that story.

It's an earful... and a mouthful!

# Maxilla Gorilla's HEADTRIP

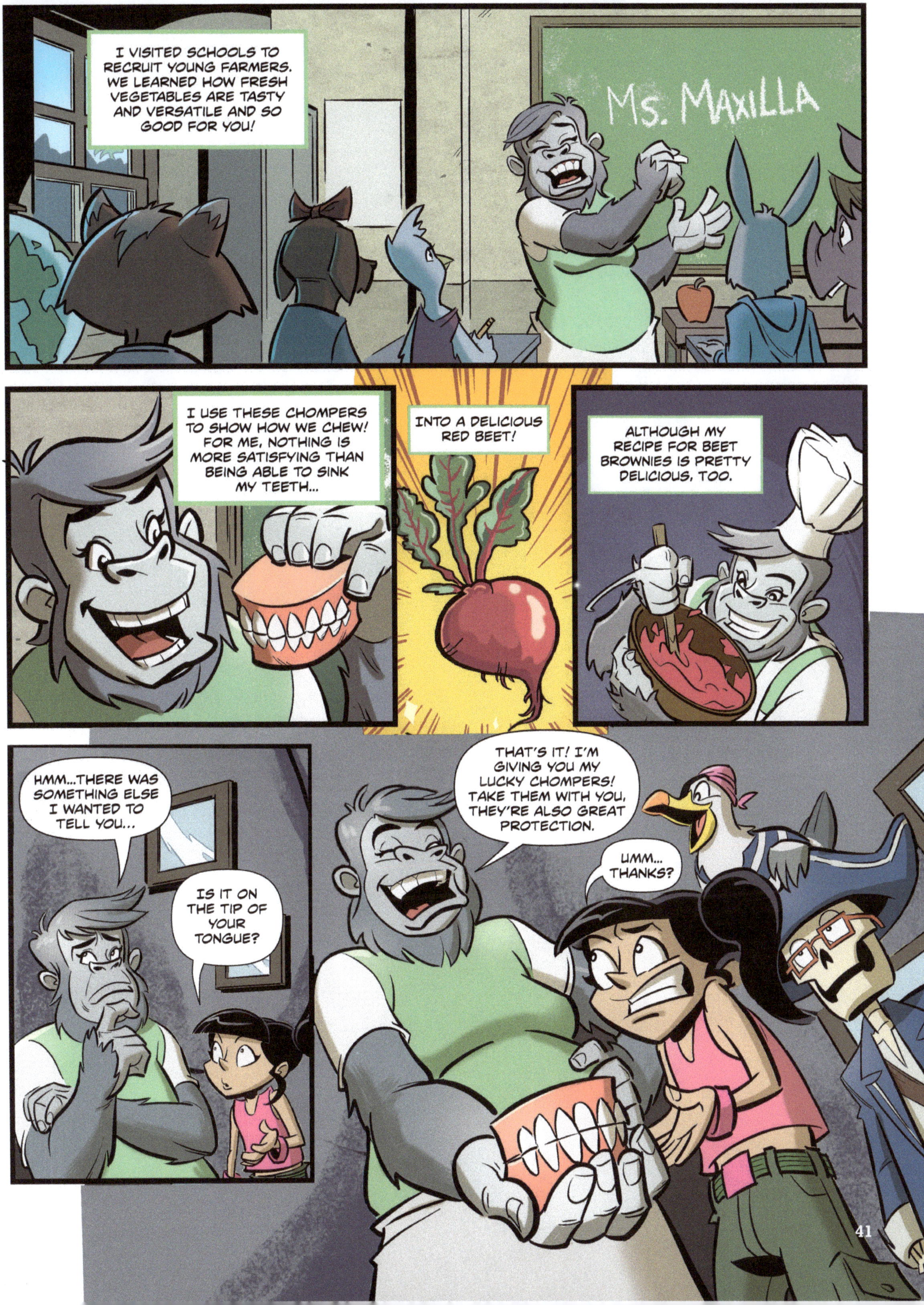

I VISITED SCHOOLS TO RECRUIT YOUNG FARMERS. WE LEARNED HOW FRESH VEGETABLES ARE TASTY AND VERSATILE AND SO GOOD FOR YOU!
MS. MAXILLA
I USE THESE CHOMPERS TO SHOW HOW WE CHEW! FOR ME, NOTHING IS MORE SATISFYING THAN BEING ABLE TO SINK MY TEETH...
INTO A DELICIOUS RED BEET!
ALTHOUGH MY RECIPE FOR BEET BROWNIES IS PRETTY DELICIOUS, TOO.
HMM...THERE WAS SOMETHING ELSE I WANTED TO TELL YOU...
IS IT ON THE TIP OF YOUR TONGUE?
THAT'S IT! I'M GIVING YOU MY LUCKY CHOMPERS! TAKE THEM WITH YOU, THEY'RE ALSO GREAT PROTECTION.
UMM... THANKS?

What's that above our heads?
That's the vomer!

# Vomer

Say it like this: "**VOE-muhr**"

The **vomer** is a long, thin bone that runs vertically through the middle of the lower nasal cavity. The lower surface of the vomer meets with the maxillae and palatine bones at the top of the hard palate. It also comes together with the ethmoid and sphenoid bones.

The vomer separates the nasal passages (the channels for airflow through the nose), helping to direct the air you breathe.

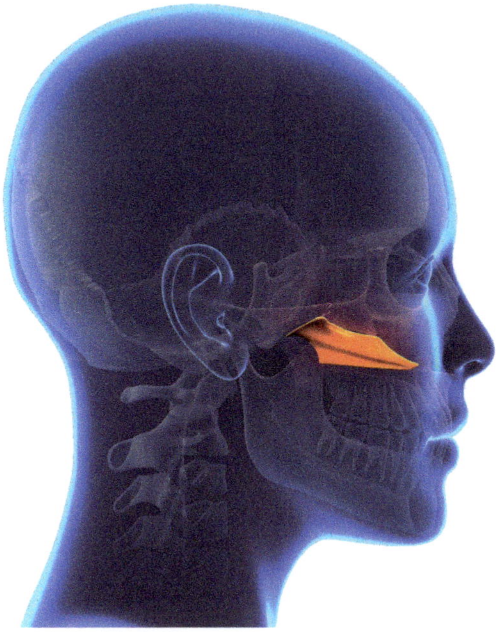

*Side View*

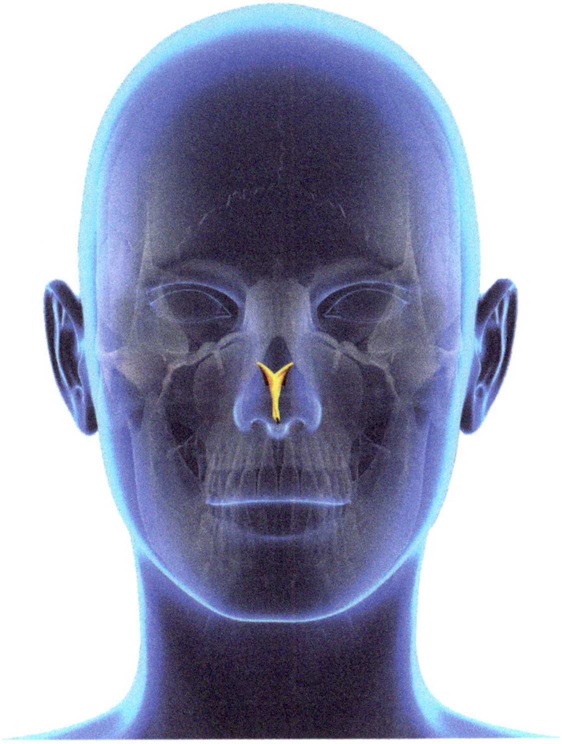

*Front View*

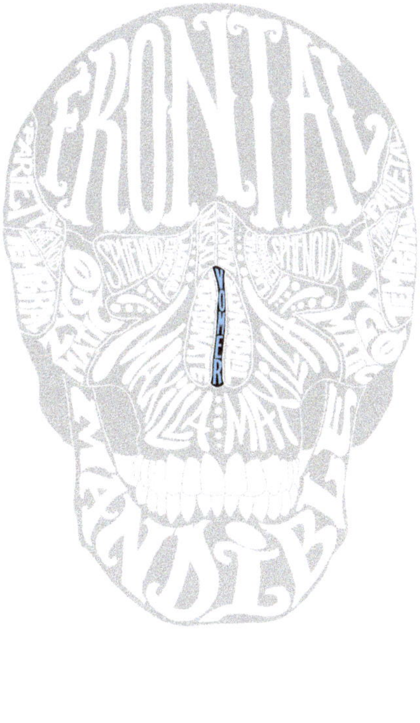

# Etymology

The word vomer comes from the Latin word that translates directly to "plowshare." A plowshare, or plow, was a long, thin tool used to dig a line in the soil so farmers could plant new seeds for the season.

*Plowshare*

Just like the plow directs the dirt, the vomer directs the air through your nose. Let's take a moment to stop and smell the roses—check out the poem below.

Place your fingers on the sides of your nose.
Press as if sneezing after smelling a rose.
The vomer bone is deep inside.
It seems like it might be trying to hide!

Though it seems thin and fragile like a glass plate,
the vomer keeps your nose chambers separate.
This bone has an important job to do:
help us breathe and smell and say a-choo!

# Activity: Pair It Up!

Remember that there are 14 bones in the viscerocranium, six pairs and two single bones.

So many bones below! Circle up the paired bones and leave the rest alone.

# Activity: Chew on This! Part II

Your jaw is amazingly strong. According to the *Guinness Book of World Records*, the strongest human bite had a strength of 442 kilograms, or 975 pounds. That's approximately the weight of two large male gorillas or five full-grown female gorillas!

*Can you pretend to chew like a dog or cat?*
Their jaws are made to bite and tear.

*How about a cow?*
A cow's jaw moves left to right—even more so than our jaws—because of the way they chew grass.

*How about a snake?*
A snake's mandible is attached to an elastic ligament, which lets the jaw stretch open wide for food much larger than the snake's mouth!

What other words can you make using the letters from the word palatine?

Example: NEAT

What other words can you make using the letters from the word mandible?

Example: LIMB

# Activity: Word Scramble



The bones of the mouth include the mandible, palatine bones, and maxillae. Unscramble the words to find out what these bones can do for you.

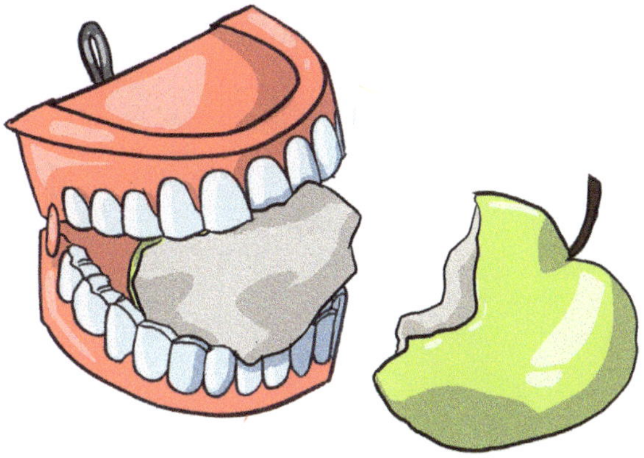

WECH

CHNMU

WGNA

TIEB

BBLEIN

CPMOH

# Activity: Skulls and Animals

The skull is nature's helmet! We can tell a lot about an animal by looking at the shape of its skull. For example, we can make educated guesses about what it eats, the animal's posture, and more. See if you can match the animal with its skull. The first one is done for you.

N
E
W
S

# Viscerocranium Review PART ONE

Use the mnemonic to help Pinky remember the first four bones of the viscerocranium. Fill in the names of each bone in the spaces below.

**M**ighty **P**inky **M**oves **V**ertically

**M** ___ ___ ___ ___ ___ ___ ___

**P** ___ ___ ___ ___ ___ ___ ___

**M** ___ ___ ___ ___ ___ ___

**V** ___ ___ ___ ___

# Mighty Pinky Moves Vertically

Mighty Mandible
Pinky Palatine
Moves Maxilla
Vertically Vomer
Are you ready to zip into the next four bones of the viscerocranium?
I'm ready to FACE IT, Dr. B! Let's go!

# Zip Lining Near Treasure

Zip – Zygomatic

Lining – Lacrimal

Near – Nasal

Treasure – Turbinate

**If you like to smile, do it seven days a week!**
**Remember the zygomatic is also known as the cheek.**

# Zygomatic

Say it like this: "**zie-go-MAT-ik**"

The **zygomatic** is a paired bone that is found in the upper part of the face and makes up the most prominent part of the cheeks. It is sometimes called a zygomatic arch.

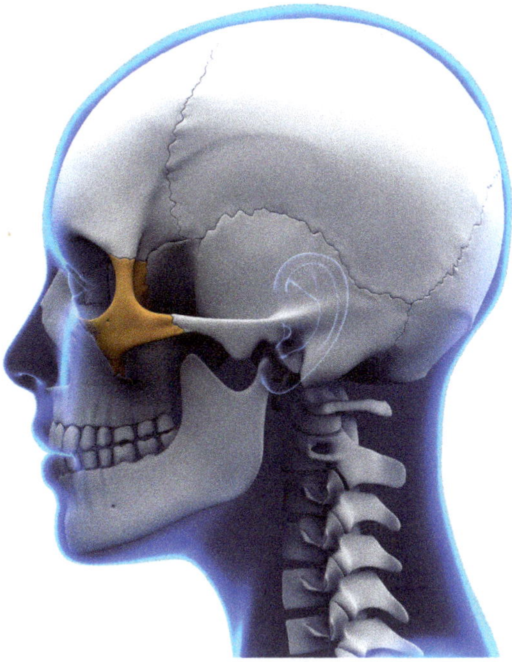

*Side View*

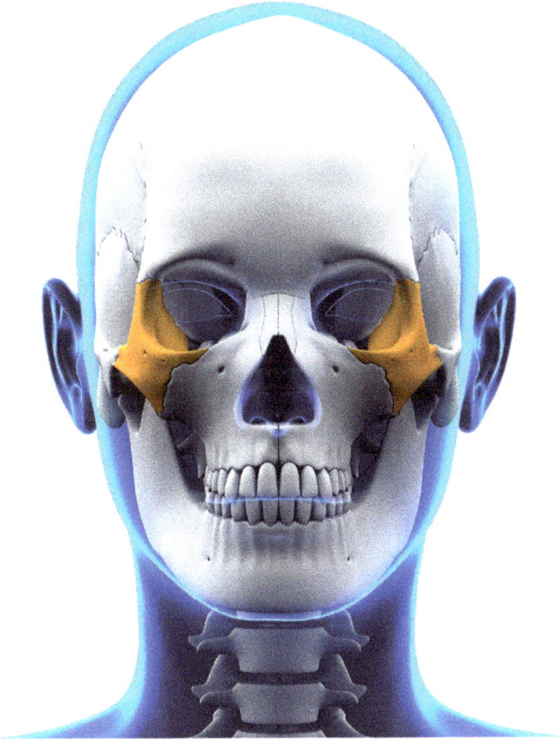

*Front View*

# Etymology

Zygomatic comes from the Greek word *zygoma*, meaning "yoke." A yoke—like the one you see here—joins two animals pulling a cart. The zygomatic bones join with the maxillae and the temporal bones of the skull. It also joins with the frontal bone and the sphenoid bones. Muscles that play important roles in chewing attach to and pass through the zygomatic bones.

# Bone Info

When someone refers to "the apples of your cheeks," they mean the highest visible point of your zygomatic.

These bones are especially noticeable on your face when you smile or laugh and are often called cheekbones.

# Activity: Face Off!

There 43 muscles in your face to help you smile, frown, laugh, smirk, and every other expression in between! But how much control do you have over your facial muscles?

Stand in front of a mirror and see if you can:

1. Squish your face like you're trying to put all of it into a straw. Then relax and do it again.
2. Raise both your eyebrows. Actually, can you raise one at a time?
3. Wink. (Start by closing your eyes. Open one eye at a time and then close it. Try both eyes! Now, try starting with both eyes open!)
4. Bring the top corners of your mouth up in a smile. That might be easy, so try bringing up one corner at a time.
5. Bring the bottom corners of your mouth down in a frown. Do one at a time.

You might find that you have more control over one side of your face than the other. Can you train your face to learn a new movement? Lift the top right part of your mouth with your finger while you hold the other side down, or try to raise the eyebrow that's hard to move while holding the other eyebrow still. It might take some time!

6. Wiggling your nose!

Although there are no muscles in your nose, you can make your nose appear to wiggle by moving the muscles around it.

Try this with a friend to train the side of your face that doesn't move as easily or to master a facial expression together. You might end up experiencing some genuine smiles... and laughter!

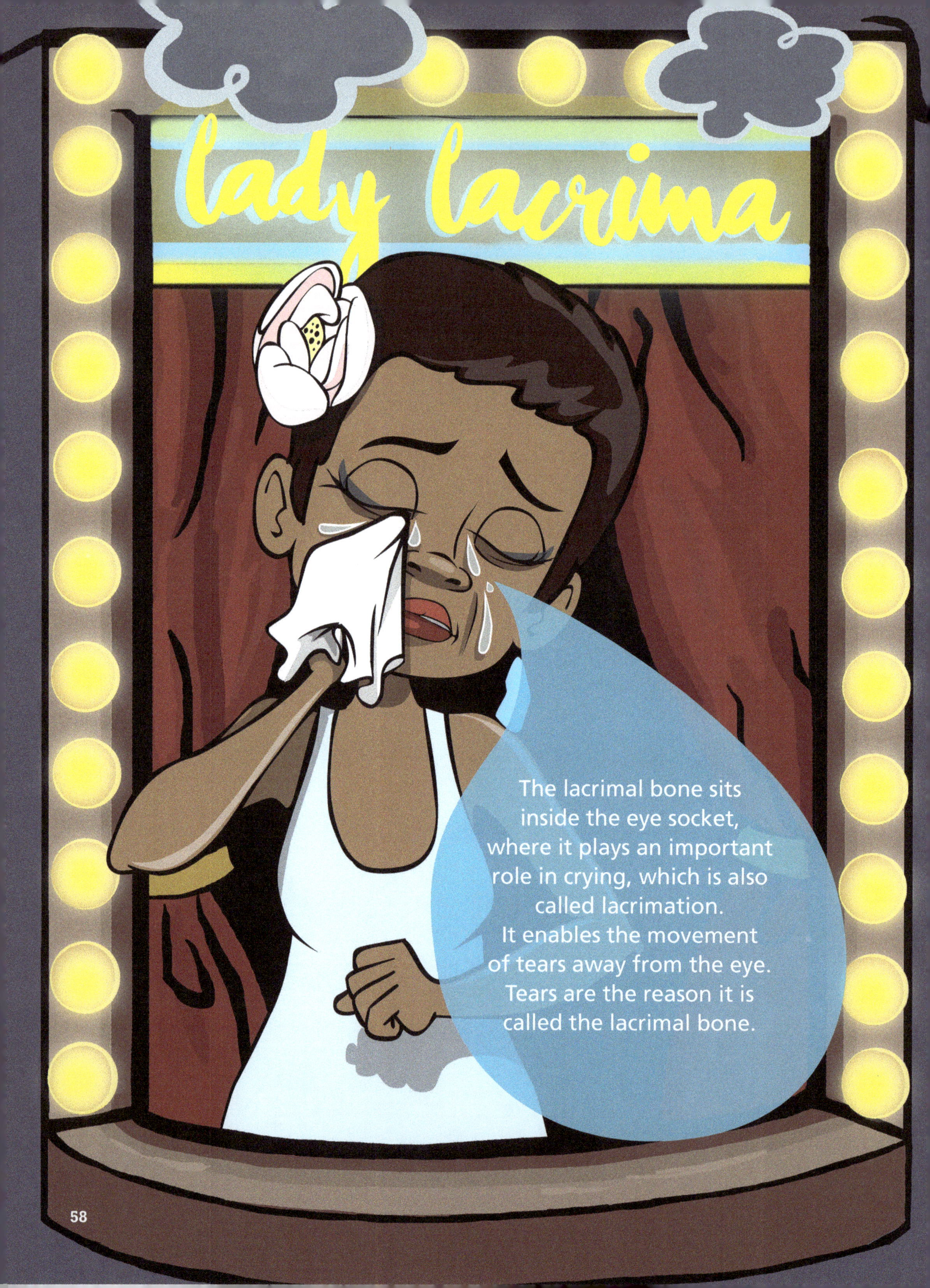
lady lacrima
The lacrimal bone sits inside the eye socket, where it plays an important role in crying, which is also called lacrimation.
It enables the movement of tears away from the eye.
Tears are the reason it is called the lacrimal bone.

# Lacrimal

Say it like this: "**LACK-rim-uhl**"

The **lacrimal** bone is the smallest and most fragile bone in the whole face!

Pinch the narrowest part of your nose, right at the top, and you will be touching the very outside edge of the lacrimal bone.

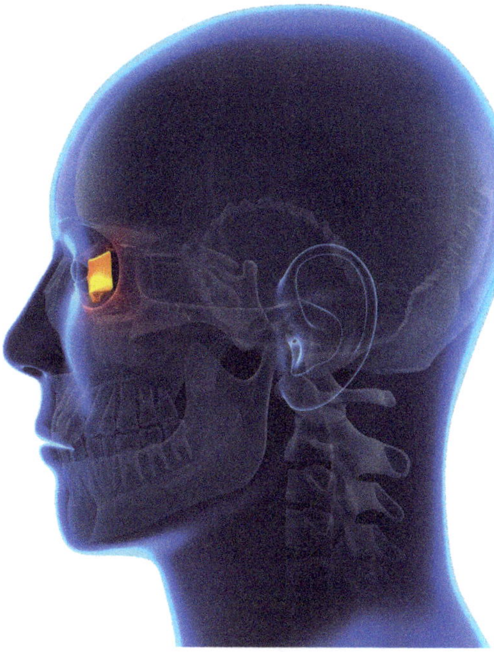

*Side View*

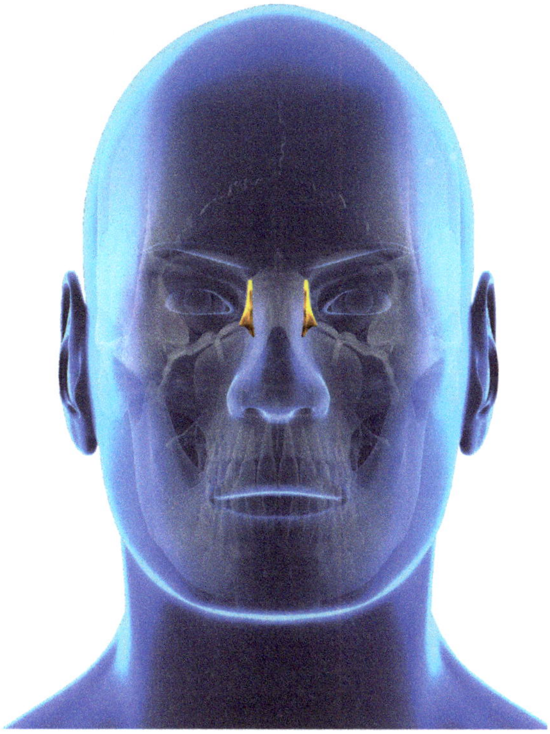

*Front View*

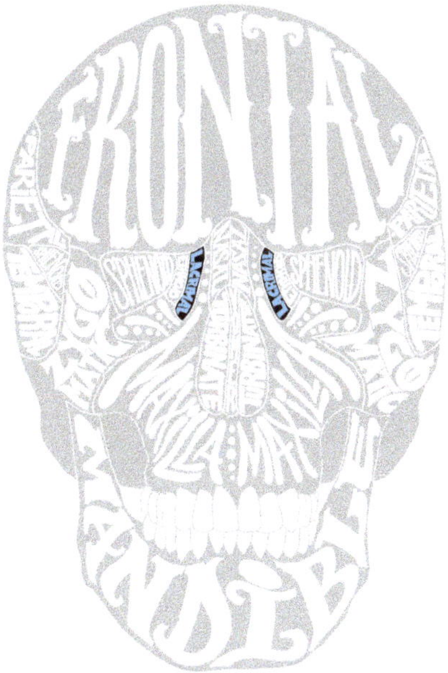

# Etymology

Lacrimal comes from the Latin word for "tear."

# Bone Info

The lacrimal bone is located just inside the eye socket.

The lacrimal bone plays an important role in the movement of tears.

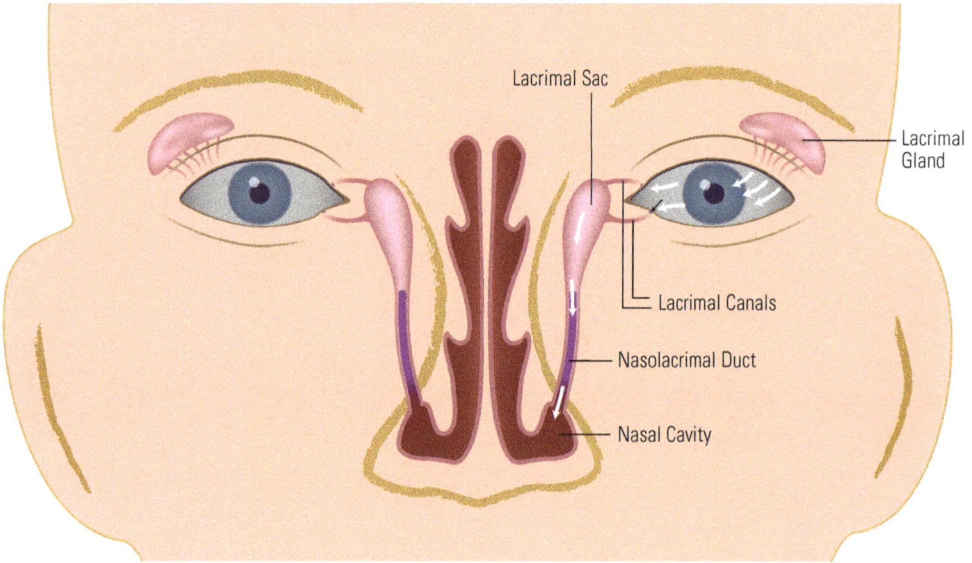


Tears are produced by the lacrimal glands in the upper eyelid. Your lacrimal glands are always producing small amounts of tears to keep eyes moist. But what happens when you produce too many tears, such as when you are crying or cutting onions?

The tears flow through the lacrimal canals, or tear ducts, in the lower corner of the eye to the lacrimal sac, which sits in a small dent in the lacrimal bone. The tears drain from the lacrimal sac into the nasal cavity. This is why you get a runny nose when you cry!

# Activity: Word Search

Try to find all the words; we helped you with the first one.

BODY
BONES
CHEW
ETYMOLOGY
FOURTEEN
GORILLA
LACRIMAL
MANDIBLE
MAXILLA
MICROVILLI
MIGHTY
MNEMONIC
MOVES
NEUROCRANIUM
PAIRED
PALATINE
PAPILLAE
PINKY
SYMMETRY
TEAR
VERTICALLY
VISCEROCRANIUM

| | | | | | | | | | | | | | | |
|---|---|---|---|---|---|---|---|---|---|---|---|---|---|---|
| L | M | L | F | Y | T | F | O | U | R | T | E | E | N | P |
| A | U | B | M | A | N | D | I | B | L | E | T | M | W | T |
| Z | I | Y | R | T | E | M | M | Y | S | Y | I | C | J | V |
| V | N | A | X | I | O | N | W | Z | M | G | M | B | K | X |
| E | A | C | L | O | O | T | A | O | H | L | I | O | G | P |
| R | R | H | T | L | G | L | L | T | T | A | C | N | O | A |
| T | C | E | P | O | I | O | Y | Y | E | M | R | E | R | L |
| I | O | W | O | A | G | X | R | P | A | I | O | S | I | A |
| C | R | I | P | Y | P | X | A | B | R | R | V | N | L | T |
| A | E | U | S | D | F | I | M | M | S | C | I | C | L | I |
| L | C | S | B | O | D | Y | L | E | C | A | L | W | A | N |
| L | S | F | L | K | J | E | V | L | O | L | L | H | U | E |
| Y | I | N | E | U | R | O | C | R | A | N | I | U | M | N |
| Y | V | I | J | M | M | T | E | B | W | E | Q | Y | X | Z |
| F | G | C | I | N | O | M | E | N | M | N | N | V | Z | N |

What's that song?
That's Lady Lana Lacrima.
Look, there she is!

Like the river meets the sea, we were meant for you and me...
lacrima

**Heads Up** *Not a single, salty seagull was harmed in the making of this book.*

# Lady Lacrima's HEADTRIP

DOOR
SORRY, LANA. WE HAD TO TAKE A GIG UP AT THE COCHLEAR CLUB.
THE LIFE OF A MUSICIAN CAN BE AS FRAGILE AS THE LACRIMAL BONE. MY BANDMEMBERS HAD TO TAKE OTHER JOBS.
DON'T CRY. LANA LACRIMA AND THE TEARDROPS MIGHT BE BREAKING UP, BUT WE WILL NEVER FORGET YOU.
TAKE THESE TO REMEMBER US BY. AND WHATEVER YOU DO—KEEP SINGING!
THESE DAYS I MOSTLY STARE IN THE MIRROR, LONGING FOR THAT OLD TEAR-DROP SOUND. I CAN'T HARMONIZE BY MYSELF!
WHAT YOU NEED IS A NEW PARTNER. AND I THINK I KNOW WHERE WE CAN FIND YOU ONE!

Do you hear that? Someone's singing!
That song sounds so familiar.
BACKSTAGE
Just like this path across the nasal bridge, you'll see our destination has a certain symmetry.

# Nasal

Say it like this: "**NAY-zuhl**"

The **nasal** is a paired bone located in the upper middle part of the face. It helps form the "bridge" of the nose.

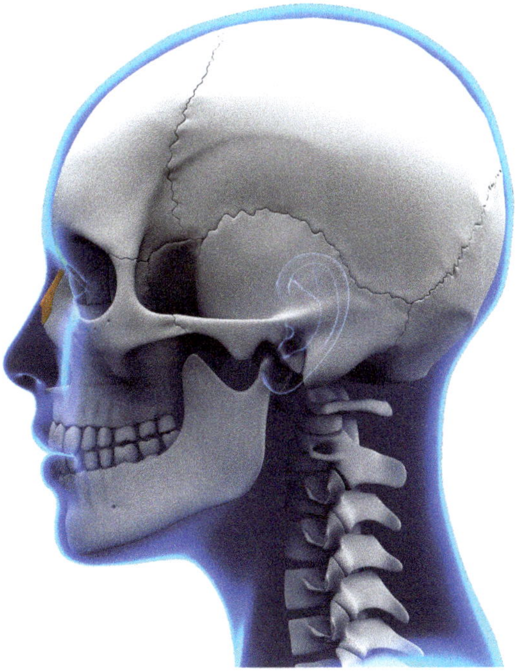

*Side View*

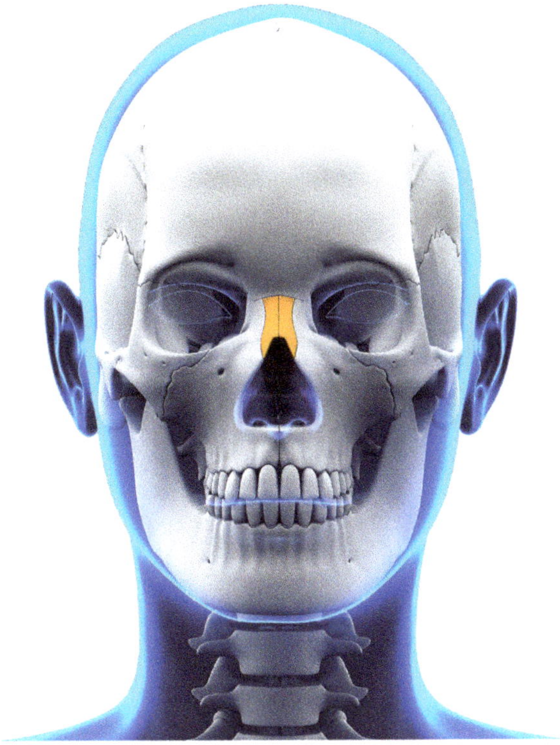

*Front View*

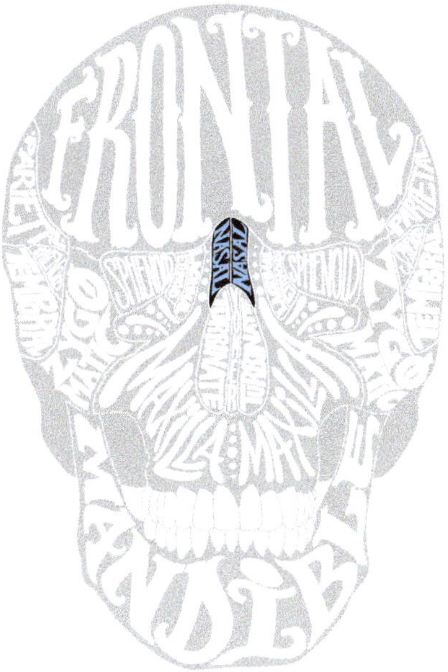

# Etymology

The word nasal comes from the Latin *nasalis*, which comes from the Latin *nasus*, or "nose." The nasal bones form the "bridge" of your nose—right where your sunglasses sit. You have two nasal bones, one on each side. They come together in the middle.

# Bone Info

The nasal bones form joints with each other and the frontal bone, ethmoid bone, and maxillae. They also connect to cartilage, a connective tissue that makes up most of your nose. Cartilage is lighter and more flexible than bone, but rigid enough to give your nose shape and to protect your nasal cavity.

Some animals, like sharks,
have entire skeletons made of cartilage.

The cartilage of a shark skeleton is much lighter than bone is, which keeps the shark from sinking to the bottom of the ocean. The flexibility of cartilage helps a shark move quickly and make sharp turns to pursue prey. The cartilage of a shark's skull and snout is denser, providing more protection of the brain and eyes.

# Activity: The Nose Knows

The nasal is one of the most unique bones in the human body. It can vary drastically in size and shape, from individual to individual.

Look at other people's faces and see how their nasal bones are different. They can be longer, shorter, wider, thinner, or even curved.

Try to draw some of the noses you see.

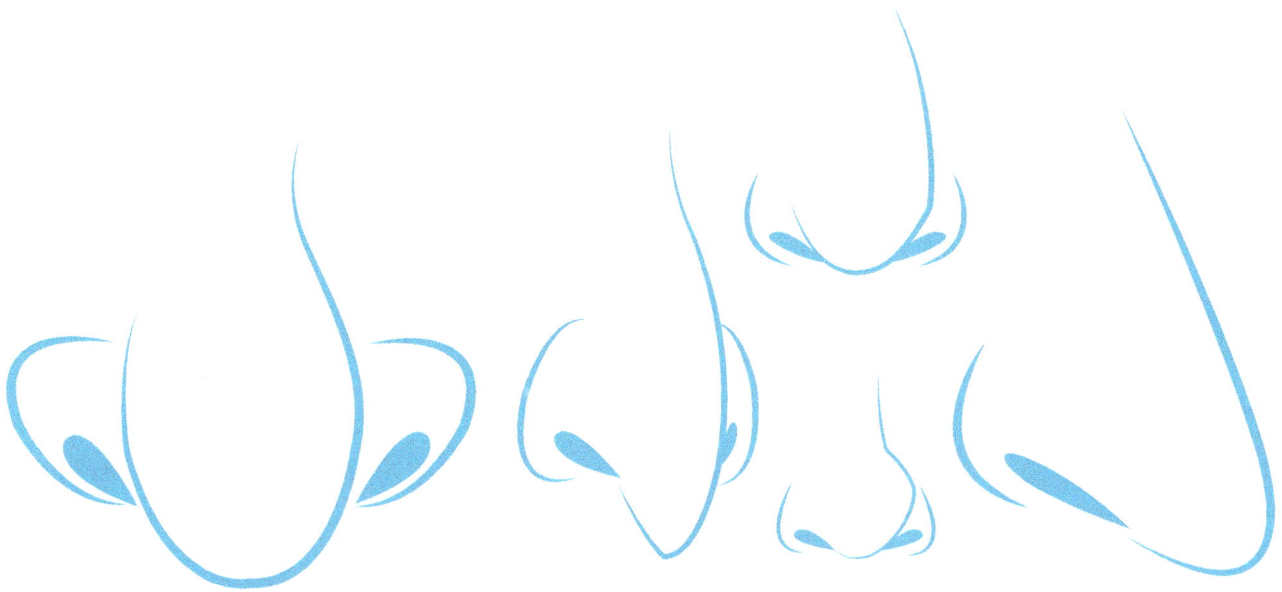

Lady Lacrima...look.
Let me see...
Like the river meets the sea...
I know that song...
Like the river meets the sea... we were meant for you...and me!
I'm Lana!
And I'm Lola!
You have a beautiful voice!
You look a lot like me!
We could be twin sisters.
With you here, I won't feel so lonely anymore.
I'm so happy, I could cry!

You guys! That's a lot of tears!
There's no need to cry anymore!
These are tears of joy! **We're quite a pair...** -sob-
Whoa! They're going to cause a flood.
Where did Seymour Seagull go?
Caw-conch!
The water level's rising. And look what just washed up to shore.
All those tears must've caused a flood. Let's head to the bones of the ear.

Hey, Herman. Can I share your shell?
No. You can't. I'm shellfish.

# Nasal Conchae (Turbinates)

Say it like this:

"**KAHN-kuh**"
"**KAHN-kee**"
"**TUHR-bin-ate**"
"**in-FEAR-ee-uhr**"
"**soo-PEER-ee-uhr**"

The inferior nasal concha is a paired bone located in the nasal cavity.

The plural of concha is conchae. The nasal conchae are also called **turbinate bones**. There are actually three kinds of nasal conchae: inferior, superior, and middle. The inferior nasal conchae are separate bones, but the superior nasal conchae and middle nasal conchae are part of the ethmoid bone. You'll learn more about the ethmoid in the neurocranium section.

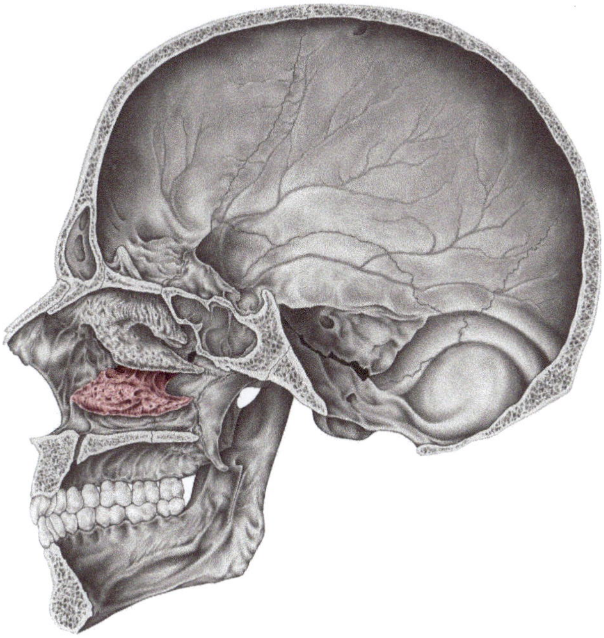

*Side View*

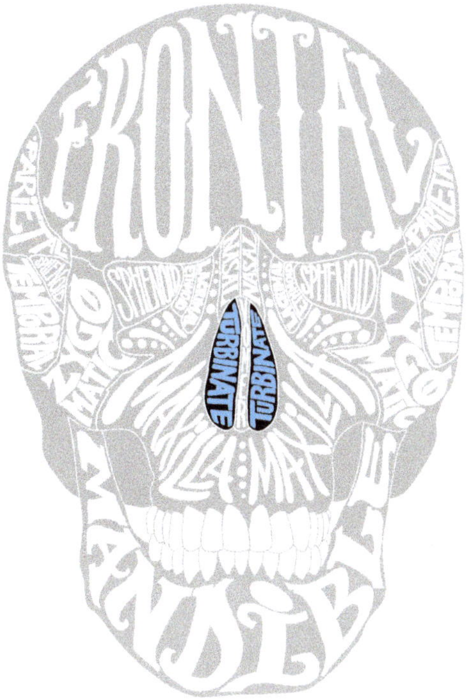

# Etymology

The word concha comes from the Latin word for "shell" or "mollusk." The Latin word is also used to name conchs, a type of marine mollusk that has a spiral shell.

Turbinate comes from the word *turbo* or *turbin*, which means "whirl" or "rotate." Think of the spinning blades of a wind turbine (also from the same root). A wind turbine produces electricity from the movement of air.

# Bone Info

The nasal conchae are thin curved bones that reach into the nasal cavity. The shape of the nasal conchae causes air to swirl as it passes through the nose, which slows the movement of air. Because air moves slower, it is warmed and humidified. Humidified air contains more moisture.

Say it like this: "**hyoo-MID-i-fied**"

The slower air movement also means more dust can be removed before air enters the throat and lungs.

# Activity: Come Up for Air

Take a deep breath, and feel the air in your nose.
Your lungs fill and mind clears as the oxygen flows.
With each and every inhale, you feel great.
Your breath is so inspiring—thank you, nasal conchae!

Fill in the blanks with the words from the bank below.
Not all words will be used.

**TURBINATES** **INFERIOR** **HUMIDIFIED**
**SWIRL** **NASAL CONCHAE** **SUPERIOR**
**AIR** **MIDDLE**

The ________________ are also known as ________________.

The ________________ nasal conchae are separate paired bones while the ________________ and ________________ nasal conchae are part of the ethmoid bone. The shape of nasal conchae make ________________ slow down, or ________________ when it enters the nose. This means air can be warmed and ________________, which adds moisture.

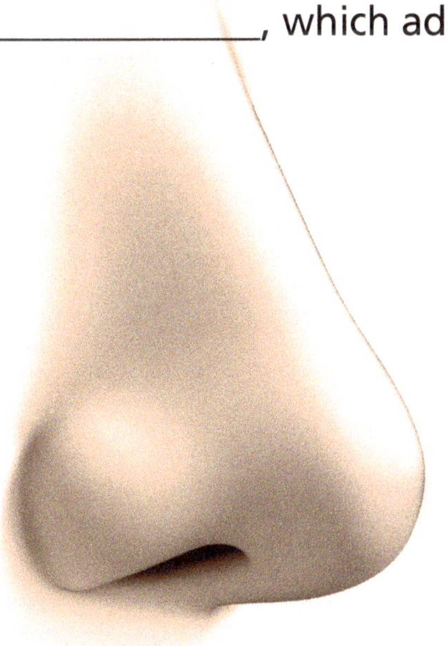

# Activity: In Your Face!

1. Which paired bone forms the bridge of your nose?

2. Which paired bone becomes more prominent when you smile?
   (Hint: It's found in your cheek.)

3. What's the name of the paired bone that forms your upper jaw?

4. Name the single bone that separates the nostrils.

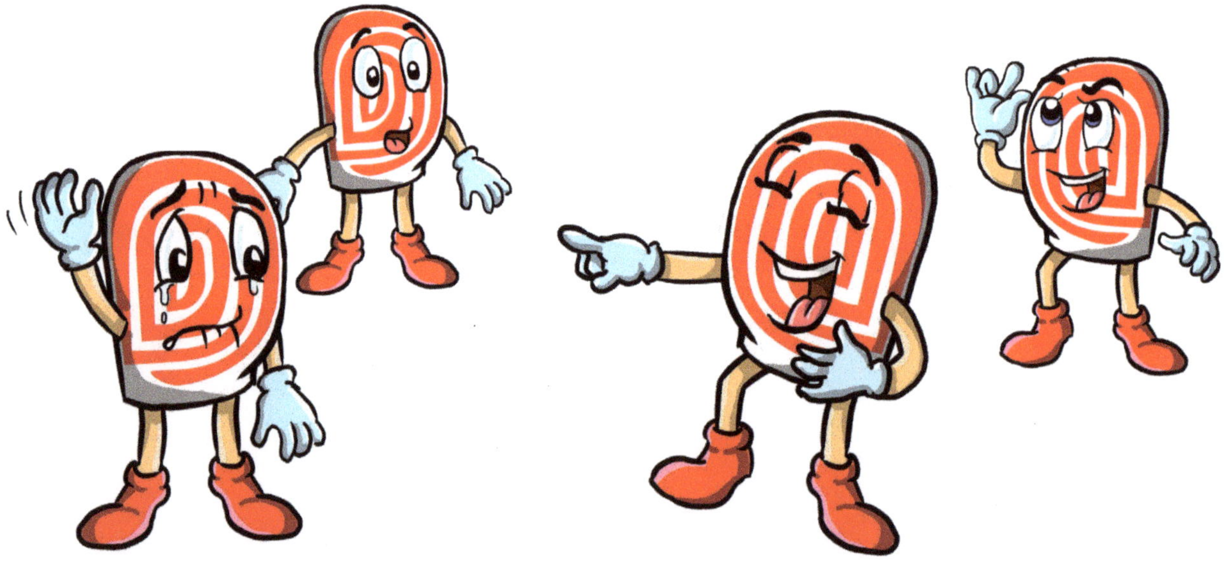

# Activity: Face Time! (LOL)

Answer the following fill-in-the-blank and short answer questions.

The viscerocranium has a total of ____ bones.

What's another name for the jawbone? ________________________________

What paired bone forms the back of the hard palate? ____________________

Which bone in the viscerocranium is shaped like a plow? ________________

What is the main function of this bone? _______________________________

Which bone is also called the cheekbone? ______________________________

Why do you think we drew Lady Lacrima crying? What are we trying to help you remember?

_______________________________________________________________

_______________________________________________________________

_______________________________________________________________

Why is another name for the nasal conchae? ___________________________

What are two things we can learn about an animal by looking at its skull?

_______________________________________________________________

_______________________________________________________________

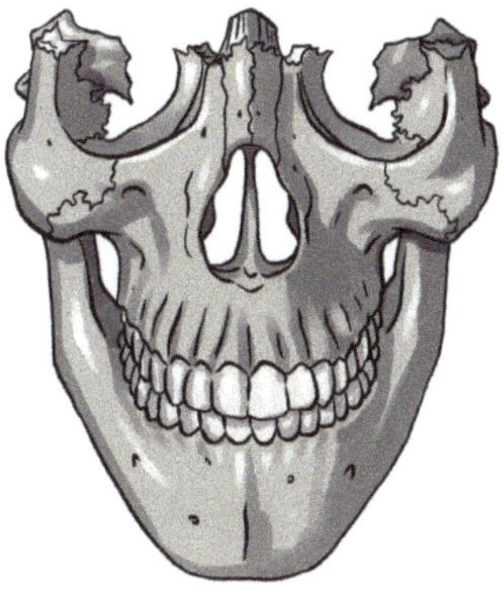

*Front View*

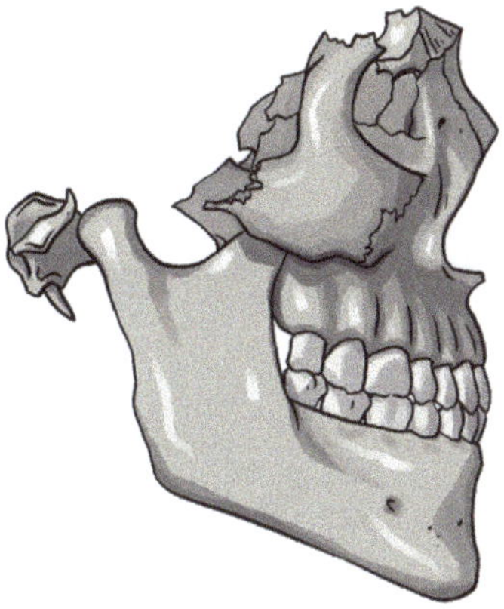

*Side View*

Now, go get a mirror. See how many bones you can identify just by looking at your reflection.

N
E
W
S

# Viscerocranium Review PART TWO

Use the mnemonic to help Pinky remember the last four bones of the viscerocranium. Fill in the names of each bone in the spaces below.

**Z**ip **L**ining **N**ear **T**reasure

**Z** ___ ___ ___ ___ ___ ___ ___ ___

**L** ___ ___ ___ ___ ___ ___ ___

**N** ___ ___ ___ ___

**T** ___ ___ ___ ___ ___ ___ ___ ___

Listening is not someone tells you

You have to listen

just hearing what
word for word.

with a heart.

— Anna Deveare Smith

I hear the ocean!
All that crying must have caused a flood! Time to go to the bones of the ears!
The ears have bones?!

# The Ossicles: Malleus, Incus, and Stapes

Say it like this: "**AWE-sick-uhls**"

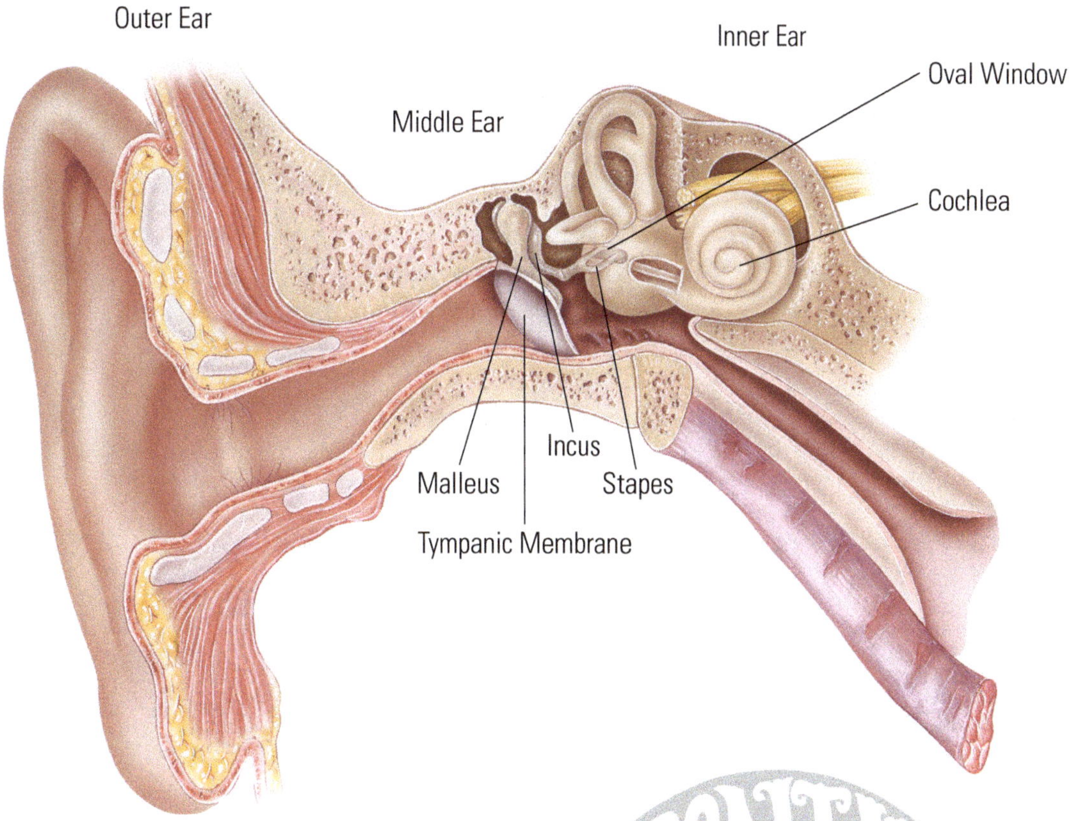


The ossicles are the smallest bones in the human body and are located in the middle ear.

Hi there!
I'm Oliver Ossicle!
Welcome!

# Etymology

The three bones of the ear are collectively called the ossicles, which means "little bones." The three ossicles are called malleus, which means "hammer"; incus, which means "anvil"; and stapes, which means "stirrup."

Say them like this:

"**IN-kuhs**"
"**STAY-peas**"
"**MAL-ee-us**"

# Bone Info

Sound enters the outer ear and vibrates the eardrum, or tympanic membrane. The membrane in turn vibrates the ossicles in the middle ear, starting with the malleus, which passes the vibrations to the incus. The vibrations in the incus are passed on to the stapes. In turn, the stapes pushes a structure called an oval window in and out, which transmits pressure waves through the fluid of the cochlea, an organ in the inner ear. Hearing receptors are located in the cochlea.

Say it like this:

"**tim-PAN-ik MEM-brane**"
"**KOH-klee-uh**"

The vibrations from loud noises can damage hearing receptors, so when you're listening to music, be sure not to blast it. If you're going somewhere noisy, consider wearing earplugs.

# Oliver Ossicle's HEADTRIP

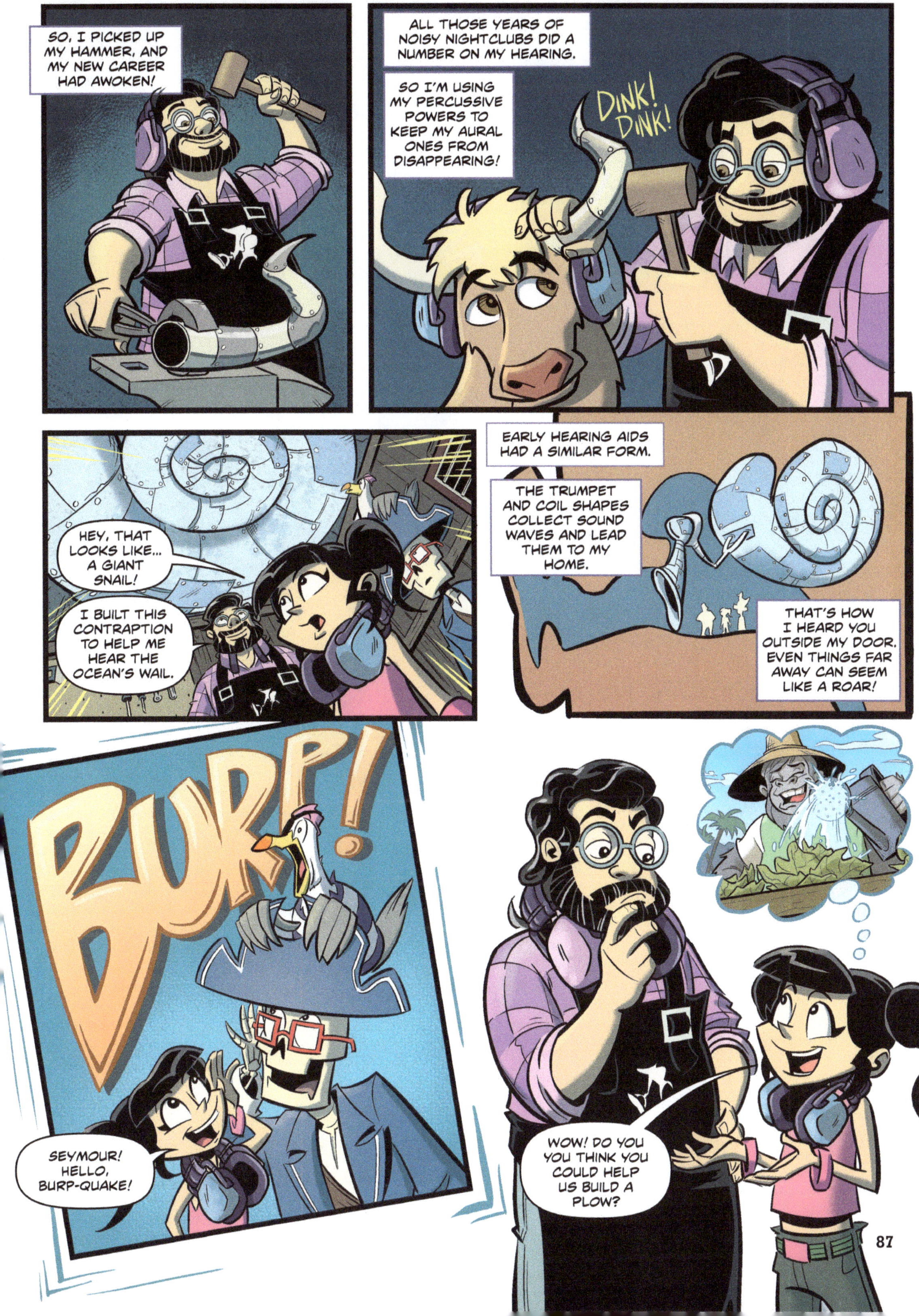
SO, I PICKED UP MY HAMMER, AND MY NEW CAREER HAD AWOKEN!
ALL THOSE YEARS OF NOISY NIGHTCLUBS DID A NUMBER ON MY HEARING.
SO I'M USING MY PERCUSSIVE POWERS TO KEEP MY AURAL ONES FROM DISAPPEARING!
DINK! DINK!
HEY, THAT LOOKS LIKE... A GIANT SNAIL!
I BUILT THIS CONTRAPTION TO HELP ME HEAR THE OCEAN'S WAIL.
EARLY HEARING AIDS HAD A SIMILAR FORM.
THE TRUMPET AND COIL SHAPES COLLECT SOUND WAVES AND LEAD THEM TO MY HOME.
THAT'S HOW I HEARD YOU OUTSIDE MY DOOR. EVEN THINGS FAR AWAY CAN SEEM LIKE A ROAR!
BURP!
SEYMOUR! HELLO, BURP-QUAKE!
WOW! DO YOU YOU THINK YOU COULD HELP US BUILD A PLOW?

# Activity: Ear Match

Fill in the missing letters, then draw a line to match each ossicle bone with its name.

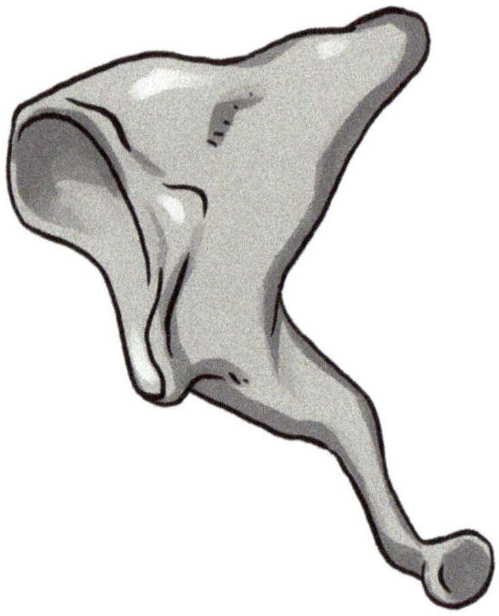

M ___ ___ L E U ___

___ N C U ___

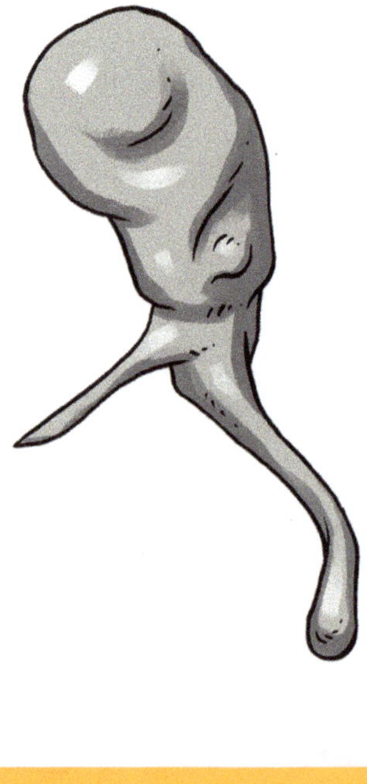

___ ___ ___ P E ___

# Activity: Do You Hear Me?

Have you ever thought about the difference in your ear when it picks up sounds both far and near?

Here's a fun activity you can do with a friend.

1. Grab some crunchy snacks (celery or a whole apple should do the trick).
2. Go to opposite corners of a room.
3. Take turns chewing as loudly as you can.

Can you hear your friend?

Can your friend hear you?

Walk toward each other until you both can hear each other.

## Try these tongue twisters on for size!

**She sells seashells on the seashore.**

**Toy boat. Toy boat. Toy boat.**

**A proper copper coffee pot.**

*During the making of Book 4, we discovered a super tongue twister:* TREASURE CHEST TEXT! *Can you say that three times fast?*

# Hyoid

Say it like this:
"**HI-oyd**"
"**LEHR-inks**"

The **hyoid** is a horseshoe-shaped bone found in the neck. It supports the tongue and **larynx**, or voice box, and is the bone to which many many muscles in the floor of the mouth attach.

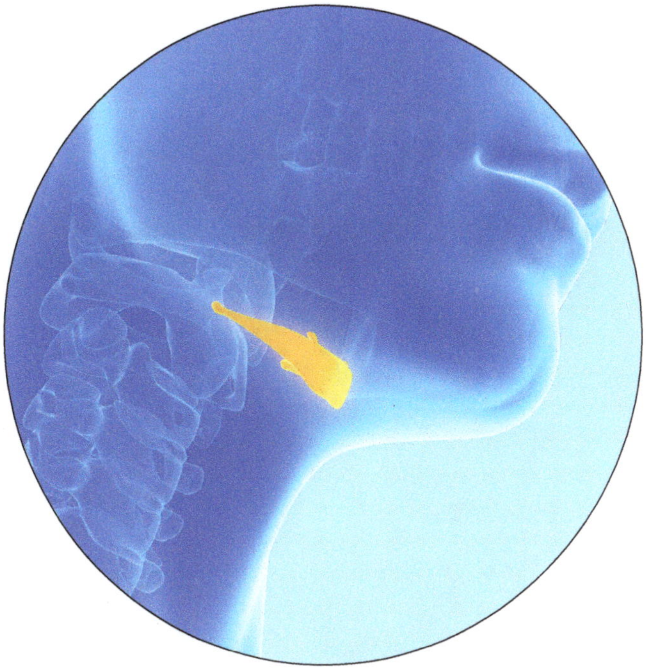

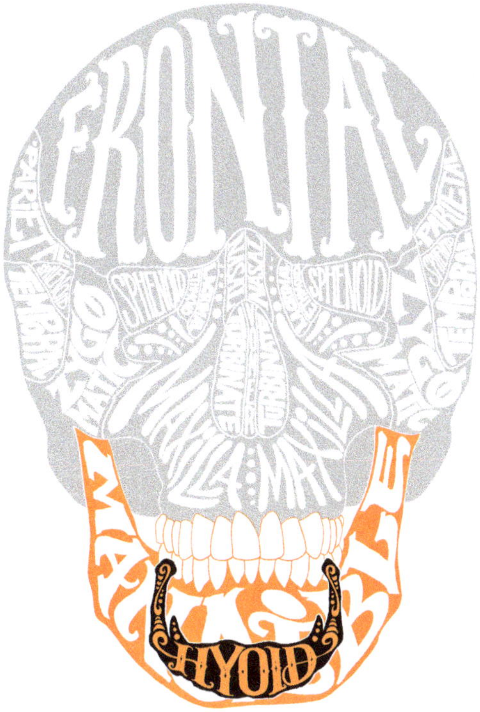


The hyoid bone is the only neck bone that's considered part of the skull. It is also the only bone in the human body that is not directly connected to any other bone. It is loosely held in place by a system of muscles and ligaments.

**Heads Up** *The hyoid is also anterior to (meaning in front of) the third cervical vertebra. Go back to Book 3 if you've forgotten where that vertebra is located.*

# Etymology

U

The hyoid gets its name from the Greek letter *huoeidēs*, which means "shaped like the letter *upsilon*" or **U**.

# Bone Info

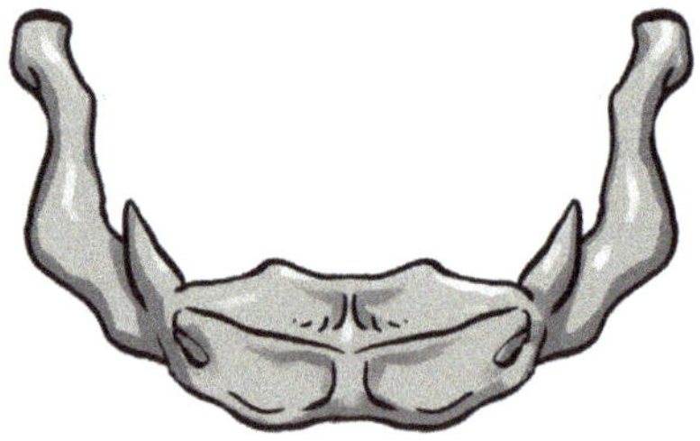

Many mammals have bones similar to a hyoid. However, only humans have a hyoid that is positioned to work with the tongue and larynx so that we are able to talk.

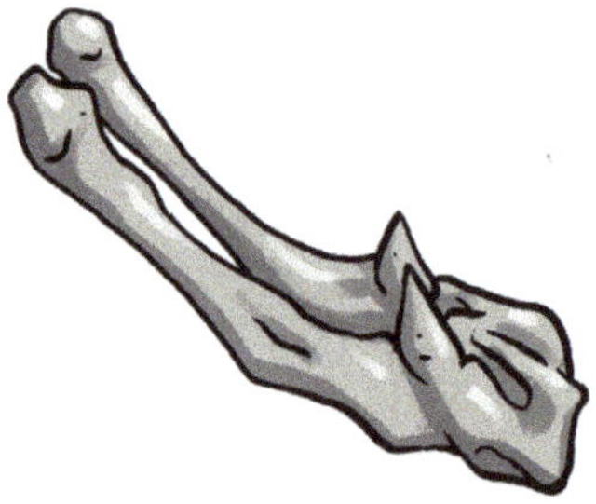

The hyoid is the main anchor point for the muscles that move the tongue. Muscles that form the floor of the mouth and play a role in moving the larynx also attach to the hyoid. Because so many muscles attach to the hyoid, it also plays a role in chewing and swallowing.

# Activity: How Well Do You Know Your Hyoid?

List three things that are unique about the human hyoid.

# The mind that opens

to a new idea
never returns to
its original size.

— Albert Einstein

# Overview of the

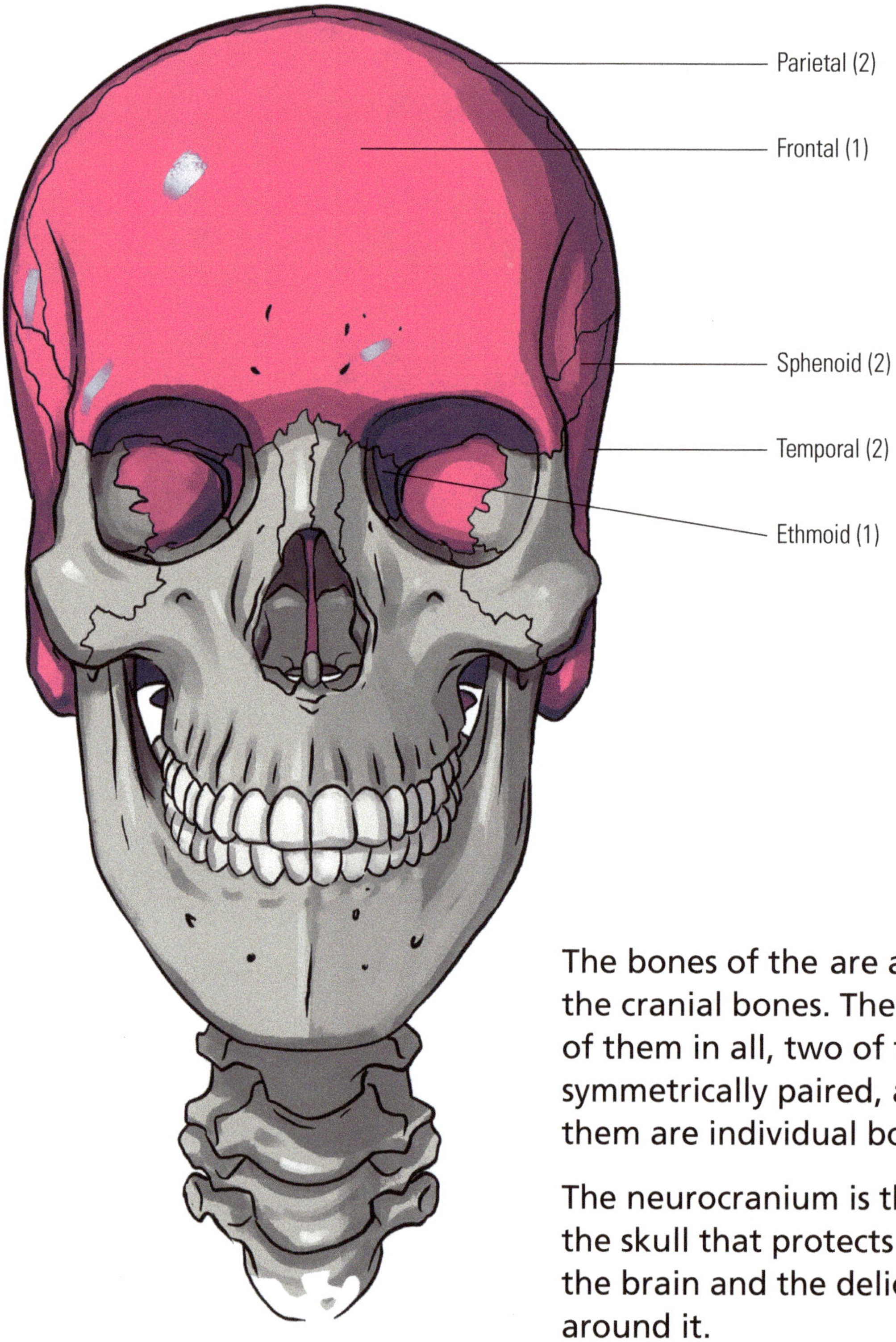


*Front View*

The bones of the are also known as the cranial bones. There are eight of them in all, two of them are symmetrically paired, and four of them are individual bones.

The neurocranium is the part of the skull that protects and houses the brain and the delicate tissues around it.

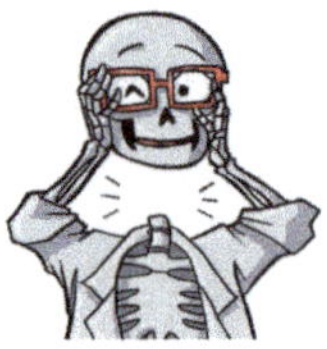

**Heads Up** *One of the individual bones, the occipital, can't be seen from this angle.*

# Neurocranium

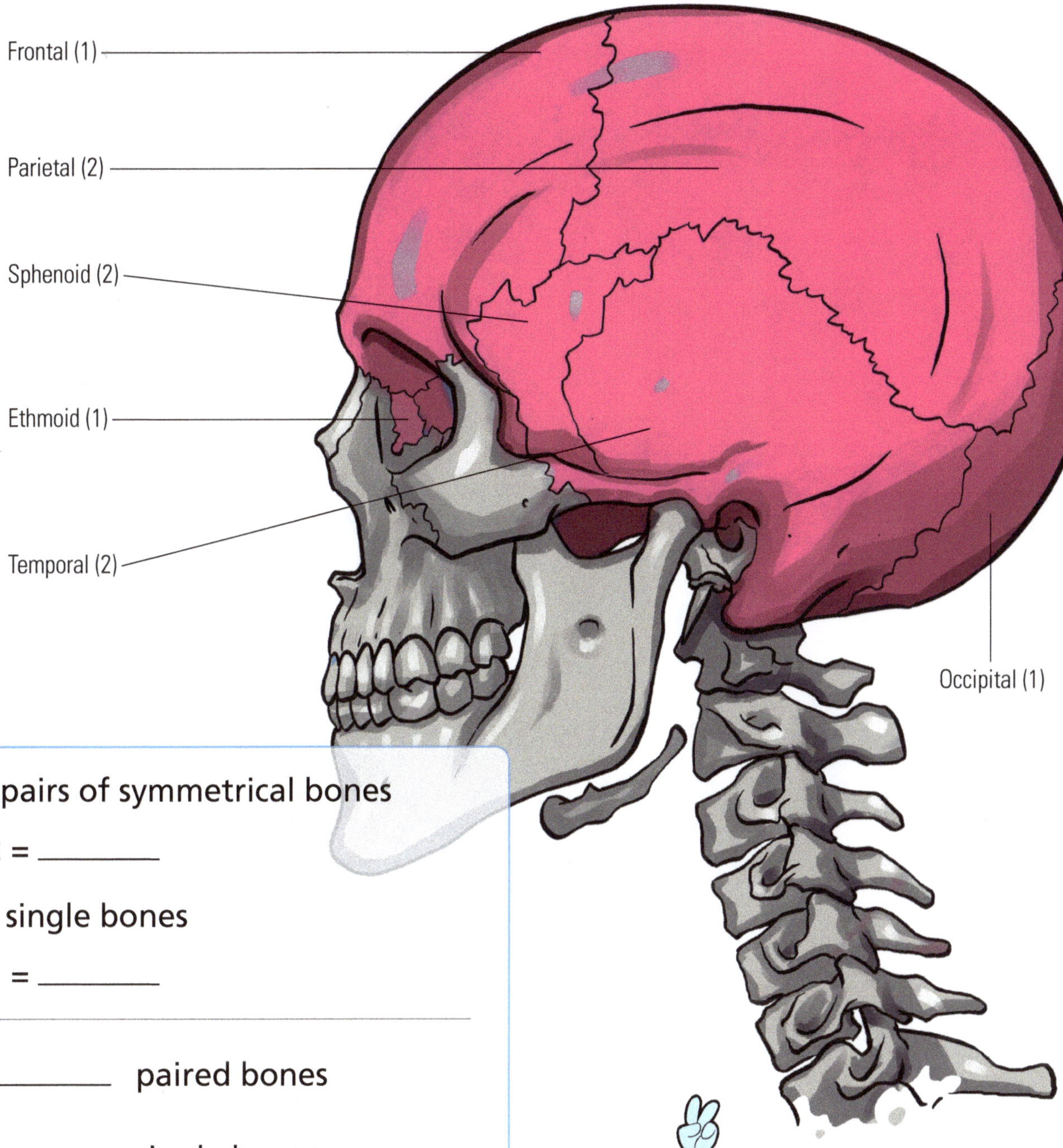


*Side View*

two pairs of symmetrical bones

2 x 2 = ______

four single bones

4 x 1 = ______

______ paired bones

\+ ______ single bones

______ neurocranium bones

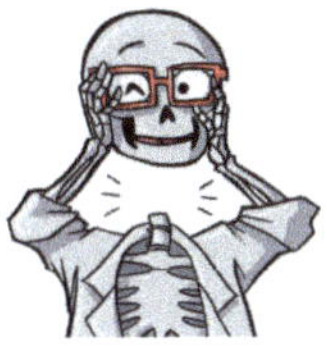

**Heads Up** *The word* neuro *is Greek for "nerves," which refers to the nervous system. The brain, which is protected by the neurocranium, is one of the main organs of the nervous system.*

We'll need a mnemonic to remember the next STEP OF the journey.
SPHENOID
TEMPORAL
ETHMOID

ARIETAL
CCIPITAL
RONTAL

Welcome to the sphenoid. You're now in the base of the neurocranium.
You can talk?!
Yes, Pinky, I'm a cartoon bird– of course I can talk! Haha!

# Sphenoid

Say it like this: "**SFEE-noyd**"

The sphenoid is a single bone that can be found "wedged" into the interior of the skull, just behind the eyes.

It helps form the base of the skull. It also helps support the eye sockets. Like other parts of the skull, it acts as a protective layer around the brain.

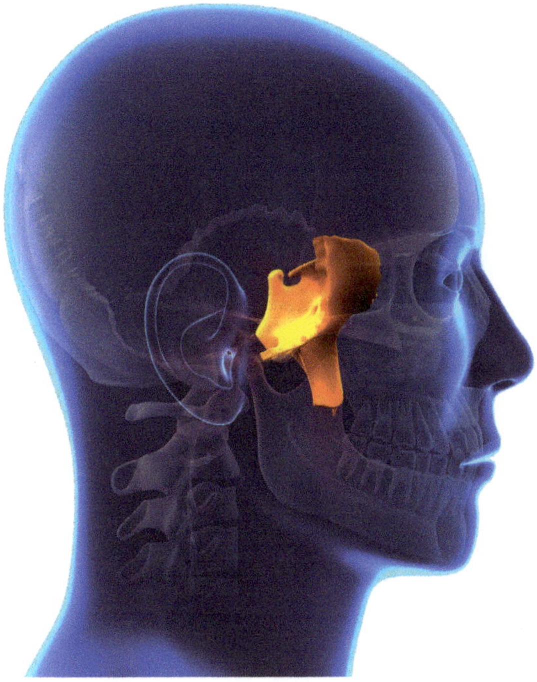

*Side View*

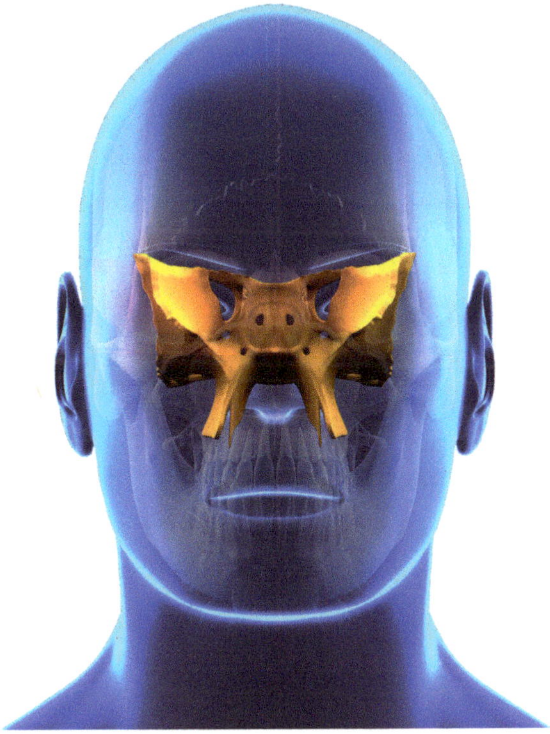

*Front View*

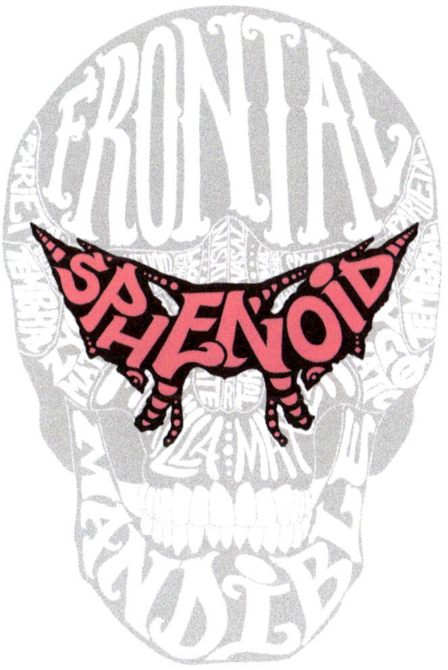


S T E P O F

# Etymology

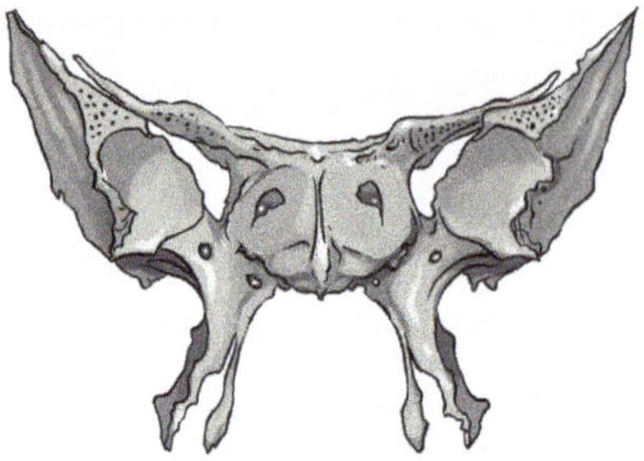

Sphenoid comes from the Greek word *sphen,* or "wedge."

# Bone Info

Because of its shape, the outer parts of the sphenoid are called "wings." Just like mine!

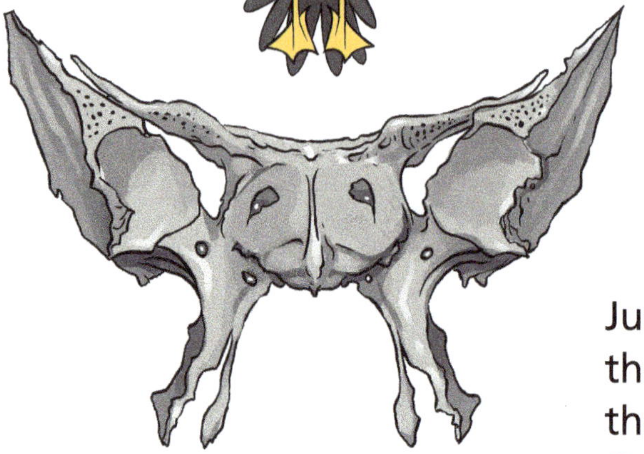

*Front View*

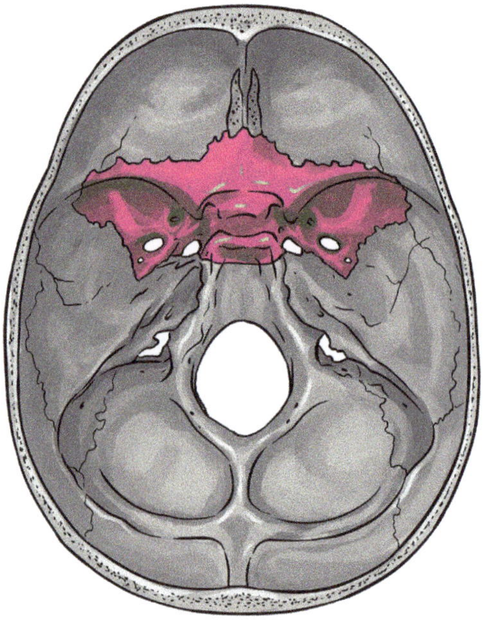

*Cross Section, Top View*

Just like the sacrum bone at the base of the spine, which you learned about in Book 3, the sphenoid bone has openings called **foramina** (singular, foramen).

There are two main foramina and multiple other structures that function as passageways for nerves and blood vessels in the head and neck.

Say it like this: "**fuh-RAW-men-uh**"

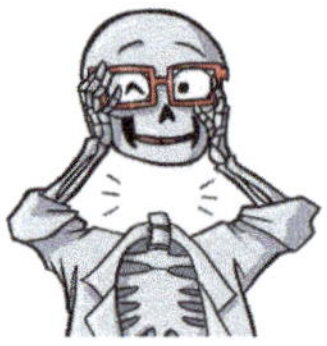

**Heads Up** *The sphenoid forms joints with nearly every bone in the skull.*

# Activity: Let's Wing It!

Sometimes wings aren't just wings on a bird. A "wing" is a word that can be used as a metaphor. A metaphor (say it like this: "**MET-uh-four**") is a word or phrase that describes something symbolically instead of literally.

Look at these examples:

1. My mom says I have to go to school so I can grow wings and fly.
2. The teacher decided to take the new student under his wing.

Can you figure out how the above things have their own kind of "symbolic" wings? Describe below.

| LITERALLY | SYMBOLICALLY |
|---|---|
| 1. Kids don't actually have wings! | |
| 2. Teachers aren't birds! | |

Can you think of anything else that has wings, in reality or as a metaphor? Write one here:

______________________________

______________________________

______________________________

# Seymour Seagull's HEADTRIP

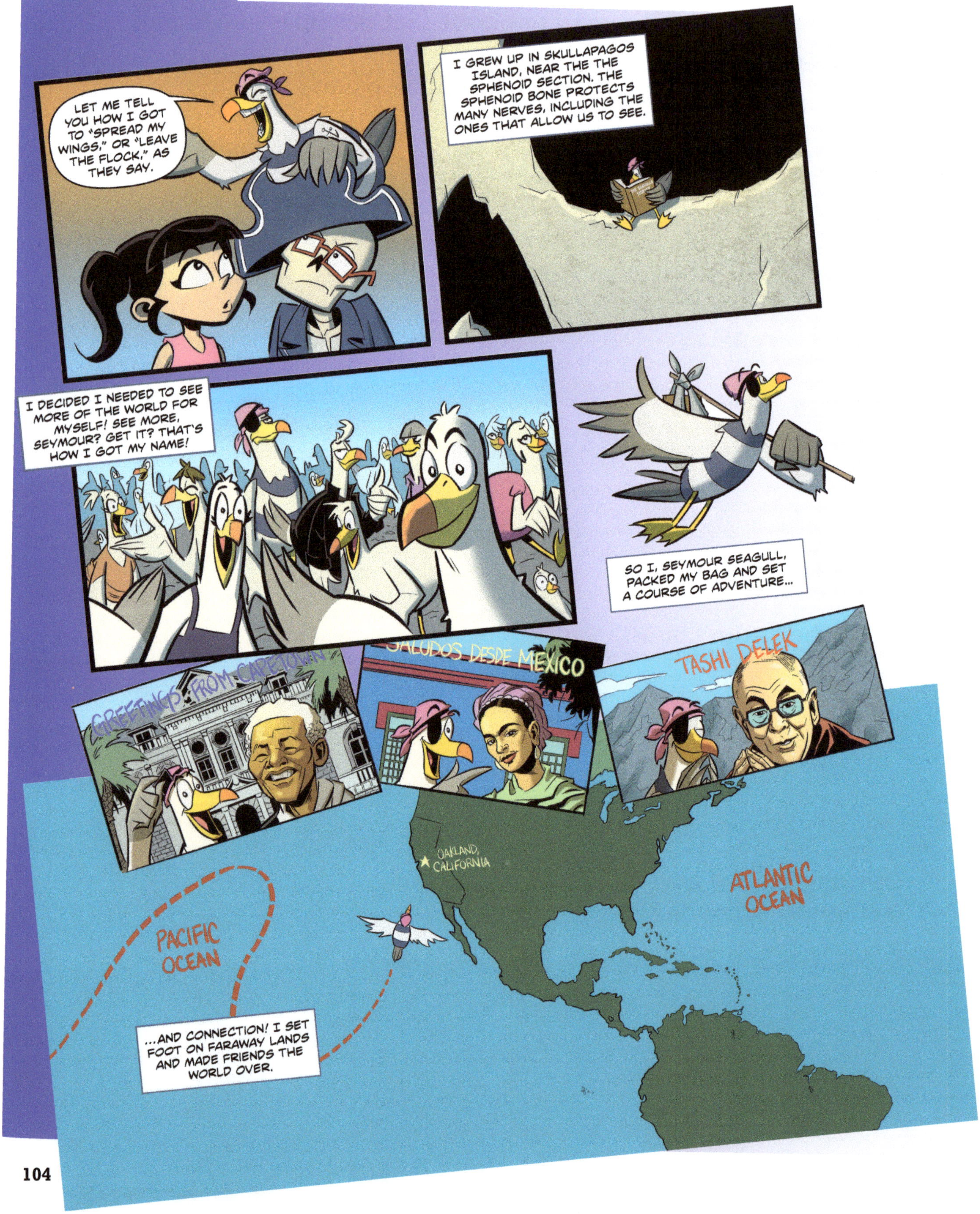

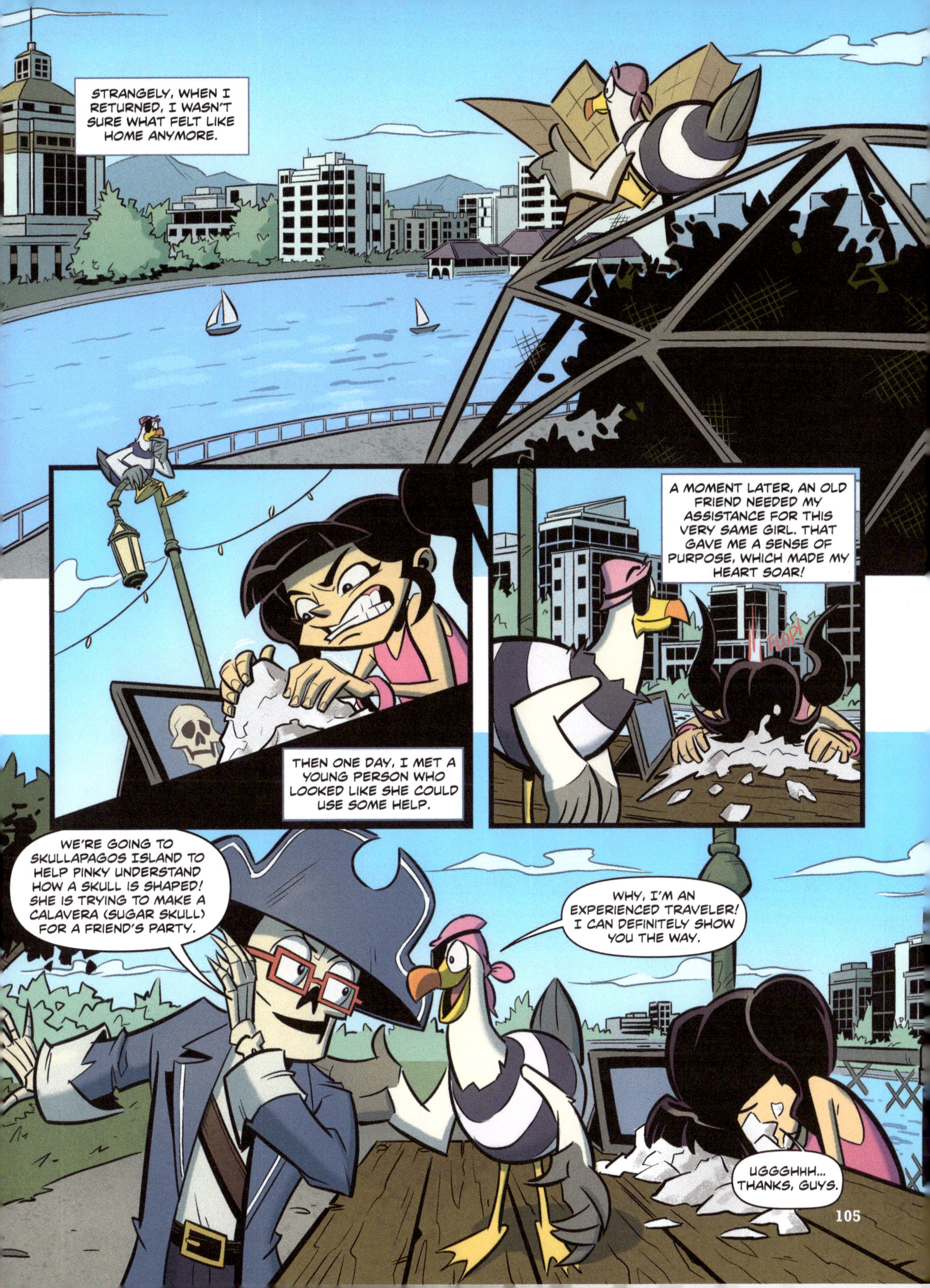
STRANGELY, WHEN I RETURNED, I WASN'T SURE WHAT FELT LIKE HOME ANYMORE.
THEN ONE DAY, I MET A YOUNG PERSON WHO LOOKED LIKE SHE COULD USE SOME HELP.
A MOMENT LATER, AN OLD FRIEND NEEDED MY ASSISTANCE FOR THIS VERY SAME GIRL. THAT GAVE ME A SENSE OF PURPOSE, WHICH MADE MY HEART SOAR!
FLOP!
WE'RE GOING TO SKULLAPAGOS ISLAND TO HELP PINKY UNDERSTAND HOW A SKULL IS SHAPED! SHE IS TRYING TO MAKE A CALAVERA (SUGAR SKULL) FOR A FRIEND'S PARTY.
WHY, I'M AN EXPERIENCED TRAVELER! I CAN DEFINITELY SHOW YOU THE WAY.
UGGGHHH... THANKS, GUYS.

# Activity: Viscerocranium? Zippity Done!

**MIGHTY PINKY MOVES VERTICALLY ZIP LINING NEAR TREASURE**

Write the names of the bones of the viscerocranium, once if it's a single bone and twice if it's a paired bone.

Record the number one or two in the column at right, identifying each bone as a pair or a single. We did the first one for you.

| Names of the bones we know so far... | | 2 or 1 ? |
|---|---|---|
| M ANDIBLE | M ________ | **1** |
| P ________ | P ________ | ____ |
| M ________ | M ________ | ____ |
| V ________ | V ________ | ____ |
| Z ________ | Z ________ | ____ |
| L ________ | L ________ | ____ |
| N ________ | N ________ | ____ |
| T ________ | T ________ | ____ |
| | | **14** |

☑ Is your sum total correct?

Stop pulling so hard. What's the hurry?
No time for falling through the cracks! I need to get to my friend Stokely's party!
Uh oh... this is definitely not normal.

We must be near the temporal bone!

# Temporal

Say it like this: "**tem-PORE-uhl**"

The **temporal** is a paired bone found behind and around the ears. It forms the lower lateral, or side, walls of the skull.

Do you remember the word "lateral" from Book 2?

If not, check back in the Book 2 glossary or look it up in the dictionary.

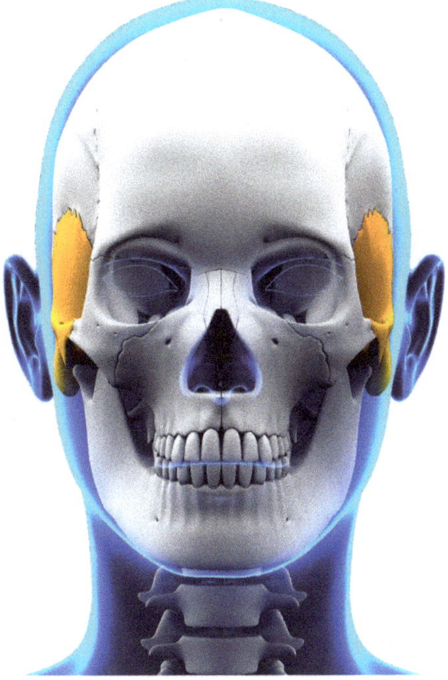

*Front View*

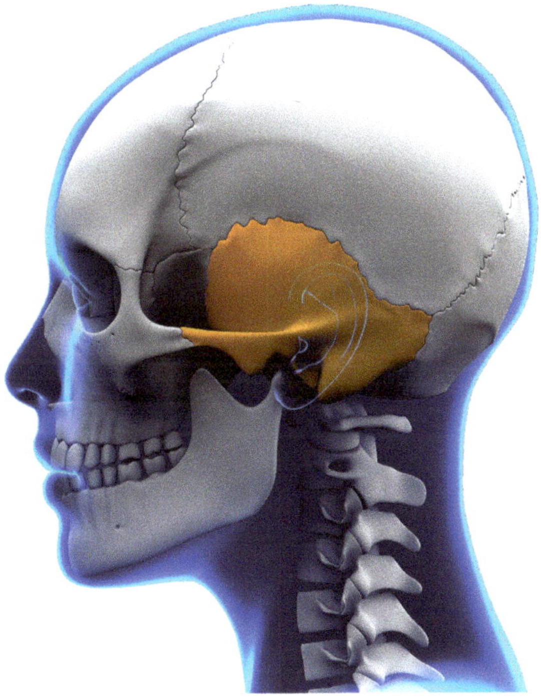

*Side View*

**Heads Up** *Remember, when we say "paired bone," we really mean two bones that are symmetrical.*

# Etymology

Temporal comes from the Latin word *tempus,* which means "time."

# Bone Info

The temporal bone protects the temporal lobe of the brain, which helps us in understanding the passage of time by maintaining long-term memory.

The temporal bones contain and protect the middle and inner portions of the ear. This section is also crossed by the majority of the cranial nerves, or the nerves that connect directly to the brain.

The lower portion of the temporal bone forms a joint with the mandible.

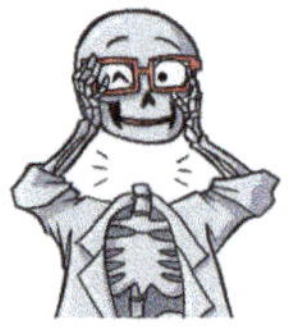

**Heads Up** *Songs have the power to trigger vivid memories that seem to transport us back in time and space. Studies have shown that for children and adults, listening to music can help to strengthen memory.*

# Activity: Who Am I?

Here are some riddles, and here are some clues. It's fun to practice the names of the bones and learning tools we use.

Fill in the blanks to answer the questions below.

1. My name has three syllables, and, all together, I make up the bones of the ear. Who am I?

   O S - S I - C L E S

2. My name has three syllables, and I'm also known as a nasal concha. Who am I?

   T _ _ - _ _ - _ _ _ E

3. My name has five syllables, and I'm also known as the cranial bones. Who am I?

   N _ _ - _ _ - _ _ _ - _ _ - _ M

4. My name has two syllables, and parts of me look like wings. Who am I?

   S _ _ _ - _ _ _ D

5. My name has three syllables, and my name comes from the Latin word for hammer. Who am I?

   M _ _ - _ _ - _ S

6. My name has two syllables, and I'm shaped like a horseshoe. Who am I?

   H _ - _ _ D

7. My name has two syllables, and I'm the smallest bone in the body. Who am I?

   S _ _ - _ _ S

I'm sinking!
Hmm...spongy...that must mean we're on the ethmoid.

# Ethmoid

Say it like this: "**ETH-moyd**"

The **ethmoid** is a square-shaped bone at the root of the nose, with many small openings for nerves to pass through.

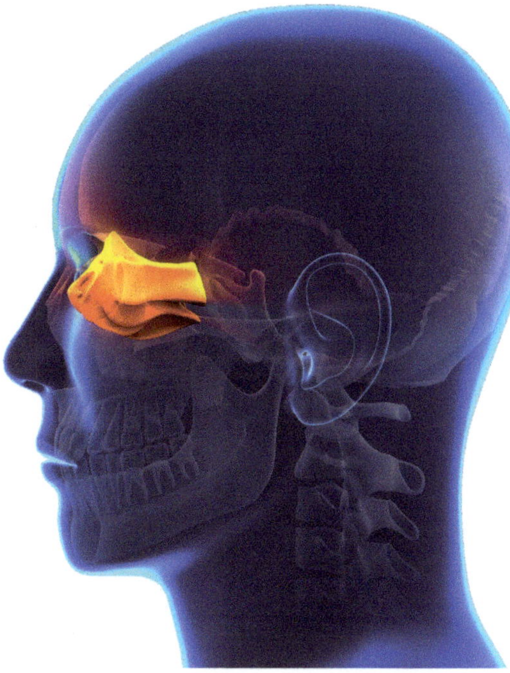

*Side View*

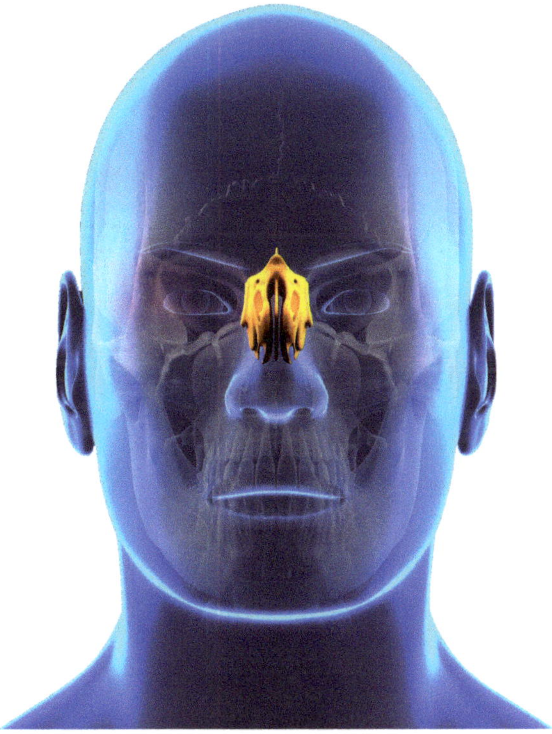

*Front View*

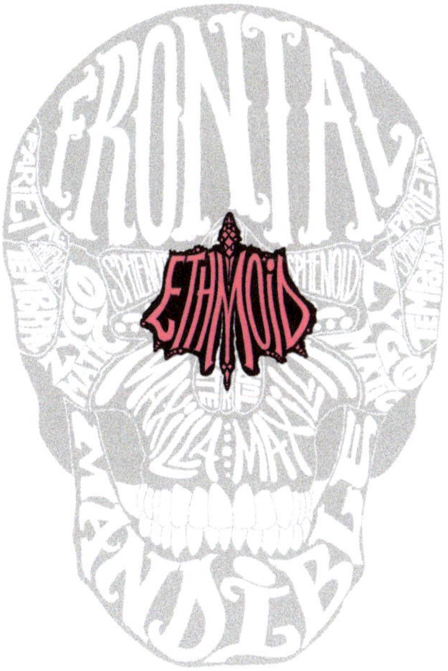


S
T
E
P
O
F

# Etymology

The word *ēthmoeidēs* is Greek for "strainer."

The ethmoid helps strain the air we breathe through our nose.

# Bone Info

From the outside, you cannot feel or see where the ethmoid is in the skull. It helps form the roof of the nasal cavity, part of the eye socket, and part of the floor of the cranium. It separates the nasal cavity from the brain.

The ethmoid has a maze of holes for all of the nerve fibers that run from the nose to the brain.

Because of its light weight and spongy structure, the ethmoid is very delicate and prone to injury.

The perpendicular plate, which helps to form the nose's middle divider, is part of the septum.

# Activity: Breath-moid

Look at Pinky's other friend Naz, below, to see how air flows into her body through the ethmoid. (She'll be going to Stokely's party, too!)

Taking slow deep breaths can calm you down, reduce stress, and improve focus. Can you think of three times a deep breath could help you? Think of a challenging situation, like feeling nervous before a test, or being angry at someone and wanting to yell.

Breathe deeply through your nose. Feel the air move into your nose, down your throat, and into your lungs. Then, breathe out through your mouth. Try this three times. Try closing your eyes while doing it.

What do you smell? What do you hear? What do you feel?
Fill in your answers below.

**I smell:**

**I hear:**

**I feel:**

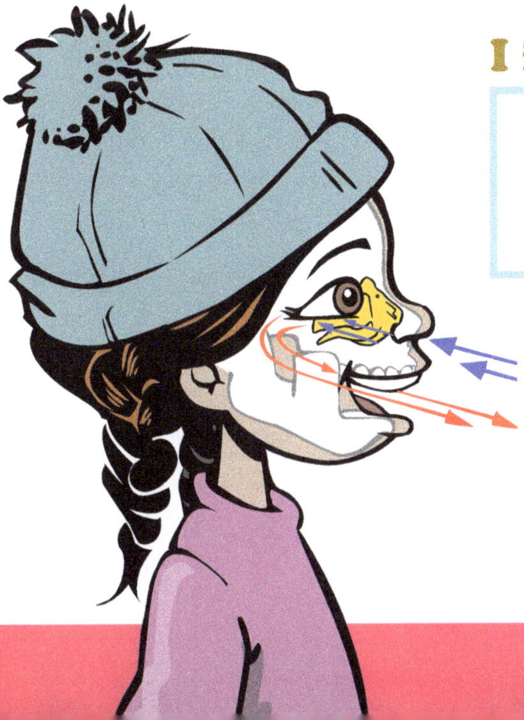

A little to the left, guys.
I thought we were going to get lost in the ethmoid! Thanks for the lift, Seymour and friends!
These spaces for nerves—which we almost fell through—make me nervous! LOL! Now off to the parietal bone.

# Parietal

Say it like this: "**puh-RYE-uh-tull**"

The **parietal** is a paired bone that is the largest bone of the cranium and connects the front and back sections of the skull.

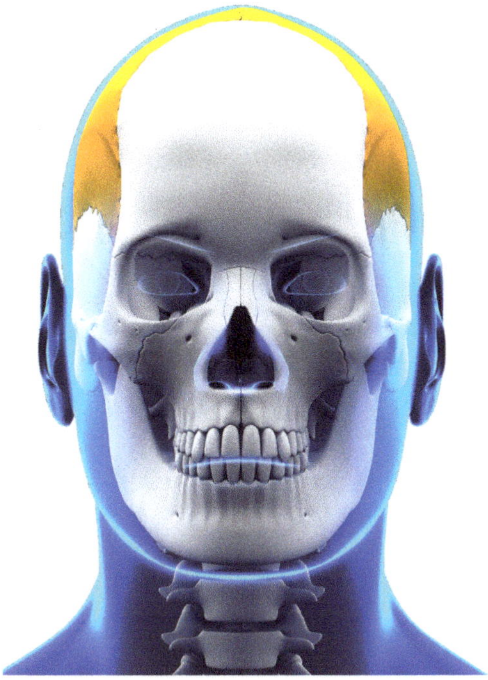

*Front View*

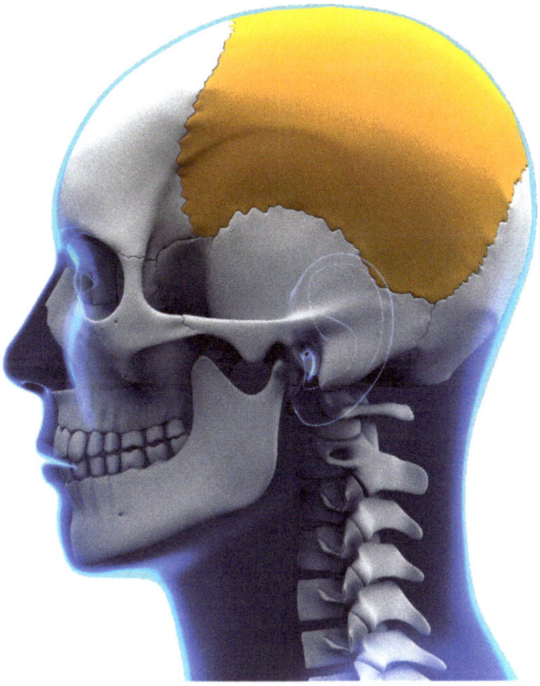

*Side View*

S
T
E
P
O
F

# Etymology

The word parietal comes from the medieval Latin word *parietalis*, which in turn comes from *paries*, a Latin word that means "wall."

# Bone Info

Can you see how the parietal bones are like a wall for the skull?

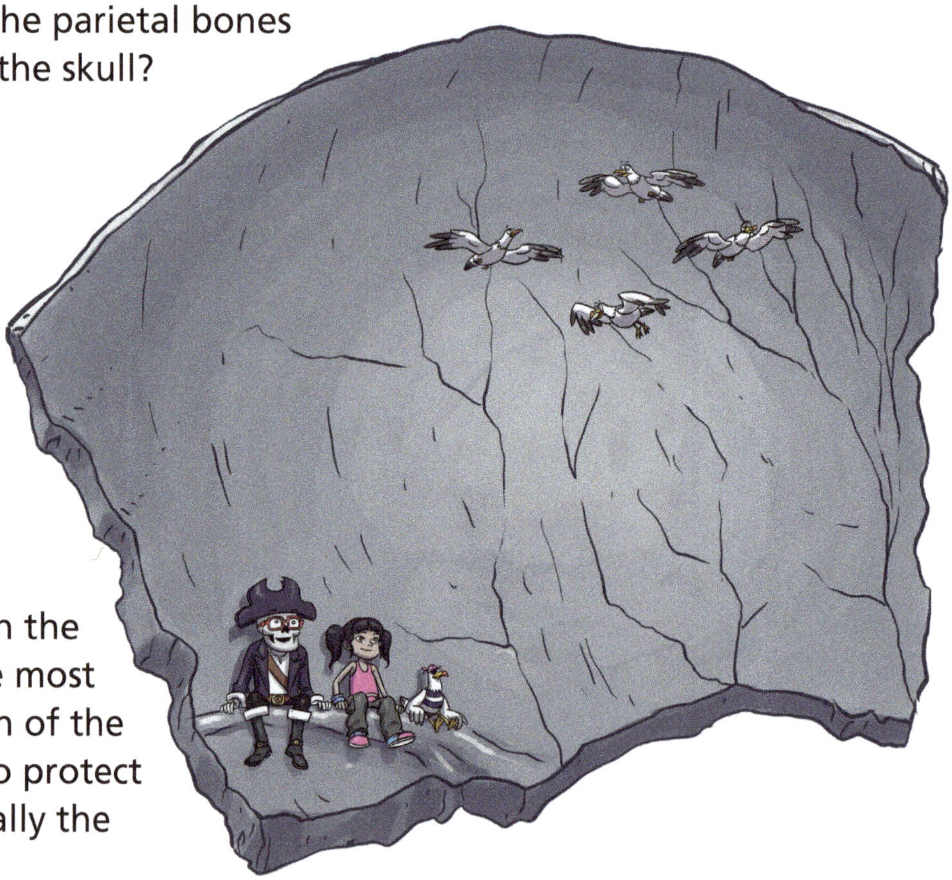

Like other bones in the neurocranium, the most important function of the parietal bones is to protect the brain, specifically the parietal lobes.

This is a part of the brain that helps us understand much of the information coming from our senses.

They also help to give shape to the head and connect to other bones such as the sphenoid and temporal bones.

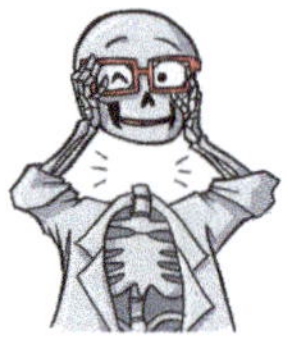

**Heads Up** *Like the temporal bones, there are two parietal bones, one on each side of the head.*

# Activity: Parietal Poise—Keep Your Head Up!

A long, long time ago, people went to "charm school" to learn manners, which included trying to develop good posture by balancing a book on their heads. Balancing a flat book on the round, dome-like shape of the parietal bone can be difficult.

Try this challenge from Pinky's friend Bounski (He'll be going to Stokely's party, too!): Can you stand up as straight as you can and balance a small book on your head? How about a soccer ball? A pillow?

As you learned earlier, the average adult head weighs ten pounds! When you get used to looking at a computer or phone, sometimes your body forgets how to hold your head upright, and the weight of your skull, brain, and so on, can strain your neck.

It's important to maintain good posture and the health of your skeleton, as well as the rest of your body.
We call that "parietal poise." Try saying that three times fast while balancing Book 4 on your head.

Dr. B...is that the ox Oliver was talking about? Look at his horn!
Yes! I believe it is, Pinky. He looks a bit lost.

# Occipital

Say it like this: "**ox-SIP-it-all**"

The **occipital** is a bone at the very back of the neurocranium. It forms part of the brain cavity, a large hollow space in the skull where the brain sits.

It's the only bone of the skull that articulates with (meaning it connects to and moves with) the cervical section of the spine.

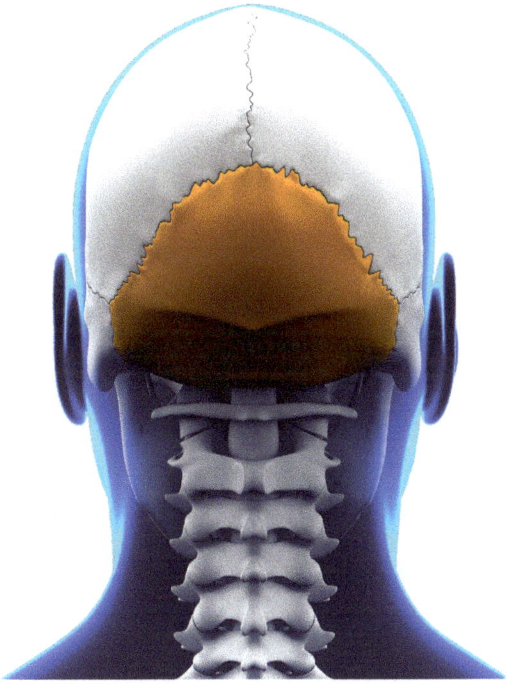

*Back View*

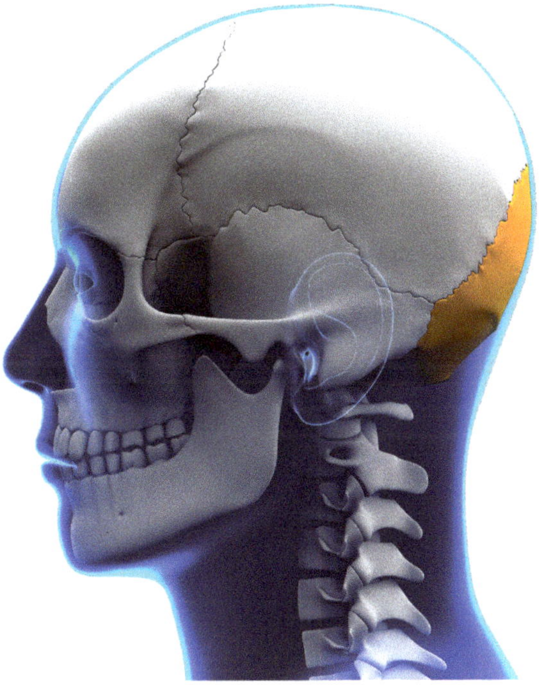

*Side View*

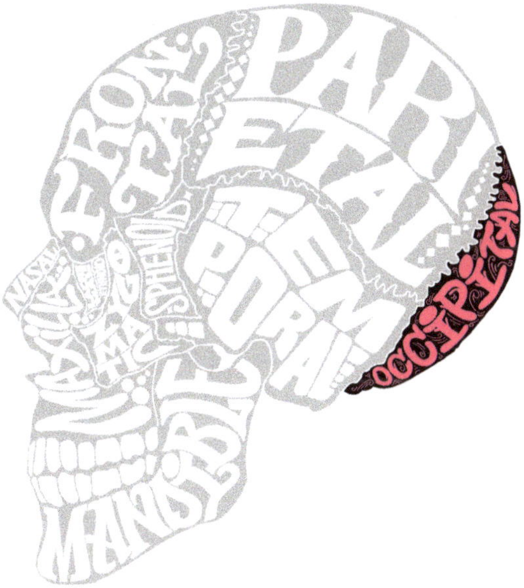


S
T
E
P
O
F

# Etymology

The word occipital comes from the Latin *occiput,* meaning "back of the skull." Occiput comes from the Latin *obs*, "against," and *caput,* meaning "head."

# Bone Info

The occipital is an important section of the skull. The occipital bone protects the occipital lobe, the part of the brain that helps process what we see with our eyes.

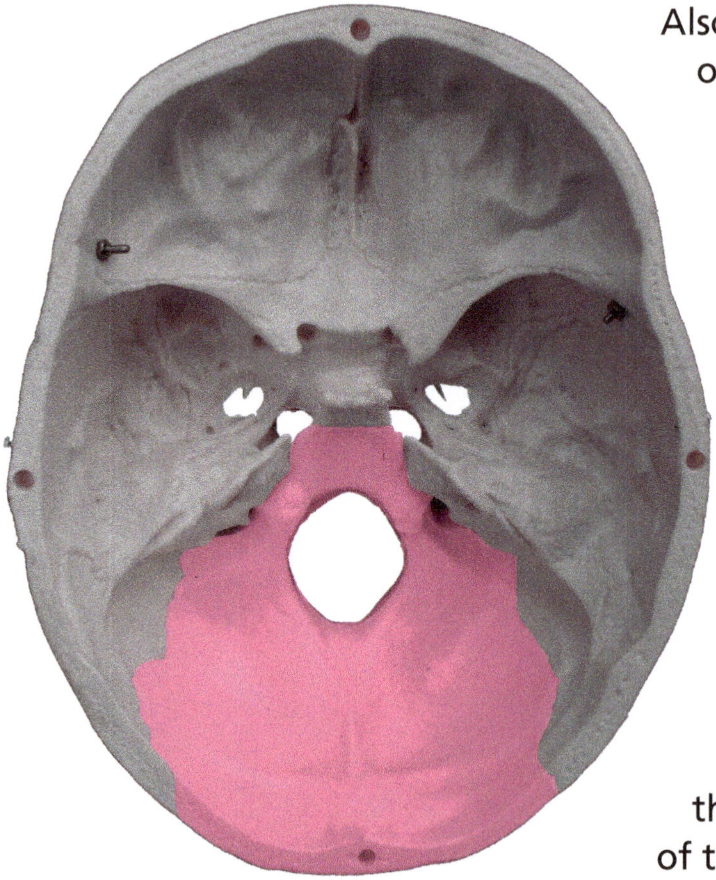

Also, this bone has a large opening in it called the *foramen magnum*. This is the opening that the spinal cord passes through so the brain can communicate with the rest of your body.

The occipital connects to the parietal and temporal bones. It is the only bone in the neurocranium that connects to and moves with the cervical, or neck, section of the spine.

# Activity: A STEP OF Knowledge

**SPHENOID TEMPORAL ETHMOID PARIETAL OCCIPITAL FRONTAL**

You've made it almost every step of the way!

Take a look at the diagram of the neurocranium. Can you label all the bones? Choose from the list above, and fill in the correct blank.

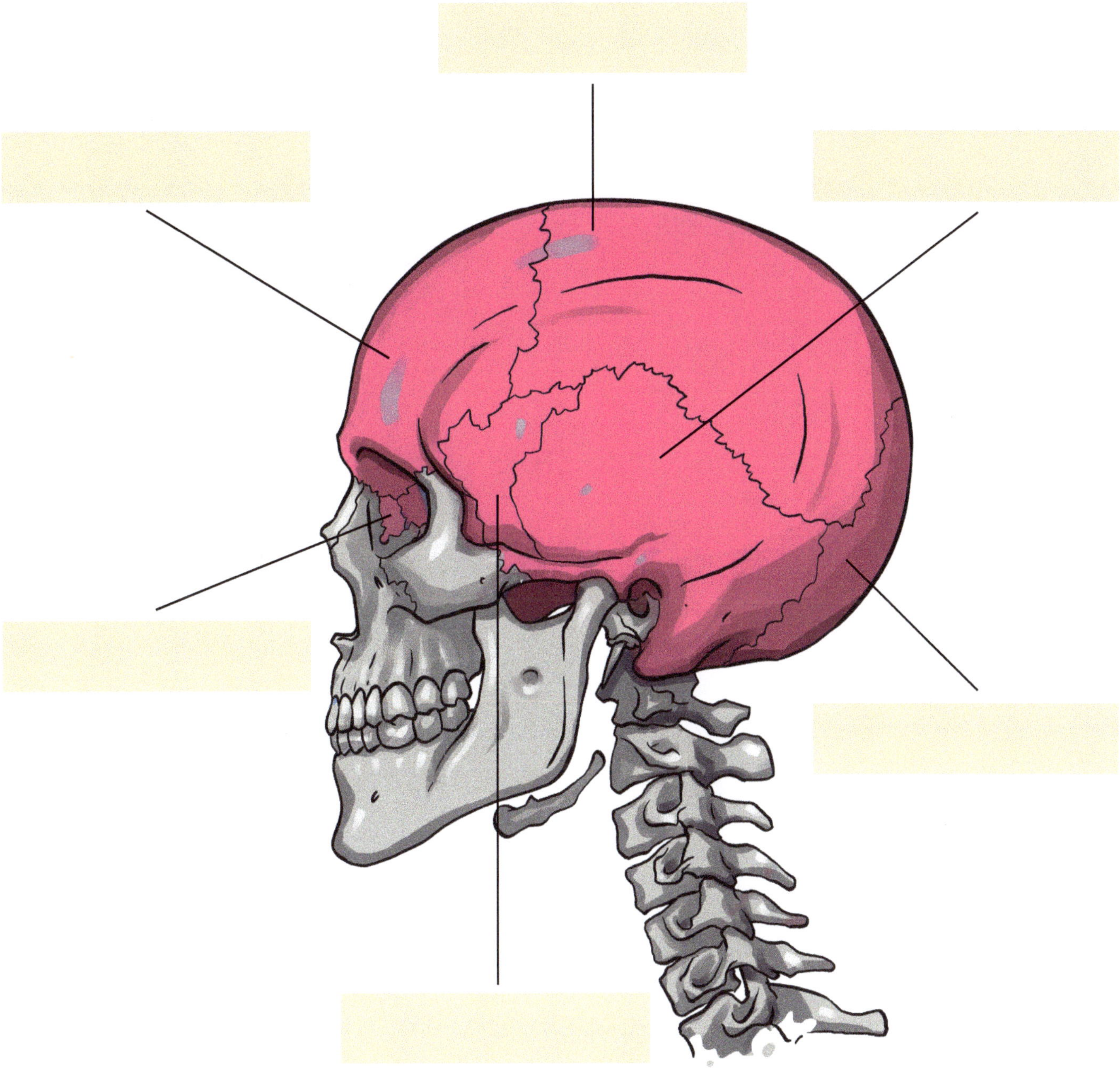

# Occipital Ox's HEADTRIP

OCCIPITAL OX FELT ALL FORLORN AS THE ONLY OX WITH UNEVEN HORNS.
TOSS!
ONE HORN HAD BROKEN AND DIDN'T GROW BACK ALL THE WAY...
...SO HE WAS LEFT OUT OF FELLOW OXEN'S PLAY.
BONK!
HA! HA! HA! HA! HA! HA!
THE OTHER OXEN PULLED PRANKS AND MADE COMMENTS UNKIND.
SHEDDING TEARS, OUR BRAVE UNI-HORN TRIED TO PAY THEM NO MIND.
AS HE FLED FROM THE MEADOW, HE FOUND A LONE BELL

WITH A BLUE COLLAR AND A WONDERFUL, SONOROUS SWELL.
DING!
THE SOUND CAUGHT THE EAR OF A METALSMITH IN HIS CAVE.
DING! DING!
HIS SKILLFUL SMITH'S HANDS FORGED A HORN-OX WAS SAVED!
NOW HIS CRANIUM BALANCED, HORNS GRAND AS A CROWN, HEAD HIGH AS HIS SPIRITS AS HE WALKED THROUGH THE TOWN.
OFT-ENRAPTURED IN PASTURE, OX SOMETIMES CLOSES HIS EYES
UNTIL HE'S WALKED A BIT TOO FAR, TO A DESTINATION SURPRISE!
PERHAPS DR. B AND PINKY CAN SEE HOW FAR HE'S ROAMED
AND OCCIPUT HIM BACK ON THE JOURNEY TOWARD HOME.

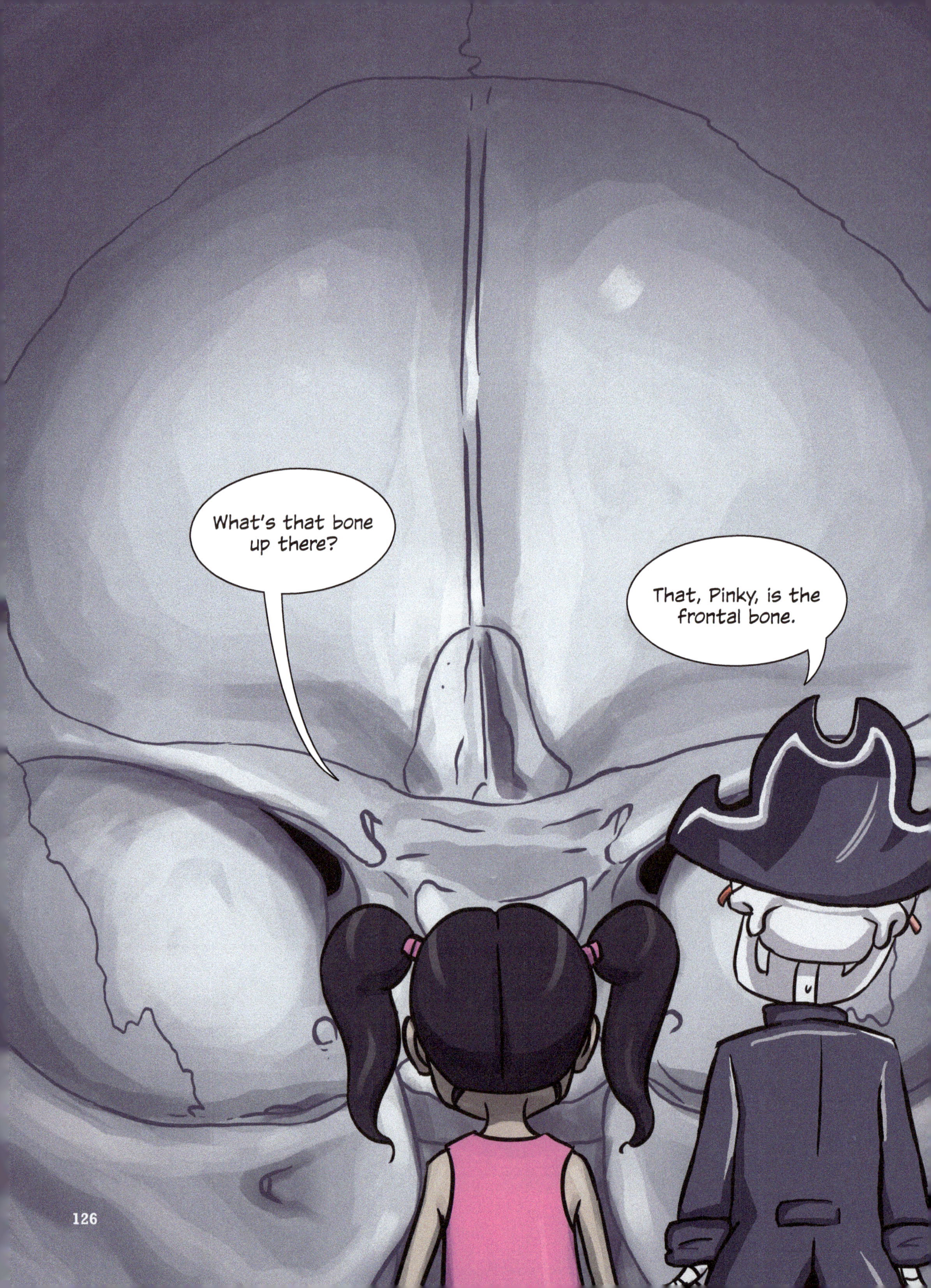
What's that bone up there?
That, Pinky, is the frontal bone.

# Frontal

Say it like this: "**FRUN-tull**"

The **frontal** is a single bone that can be found in the front of the skull. It begins just above your eyes and includes the entire forehead, forming the front of the head.

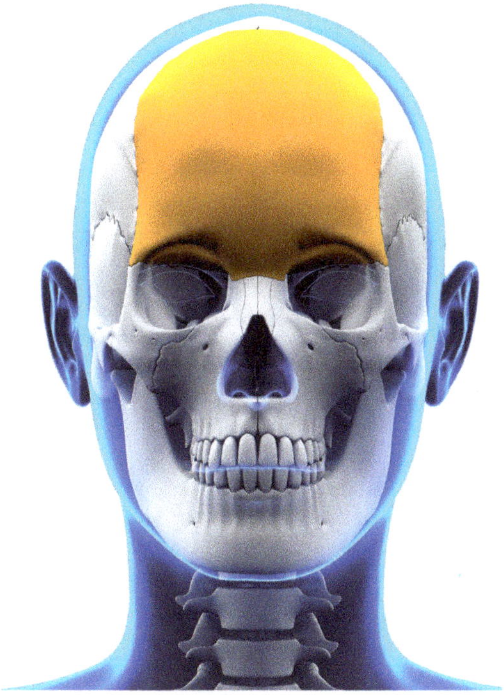

*Front View*

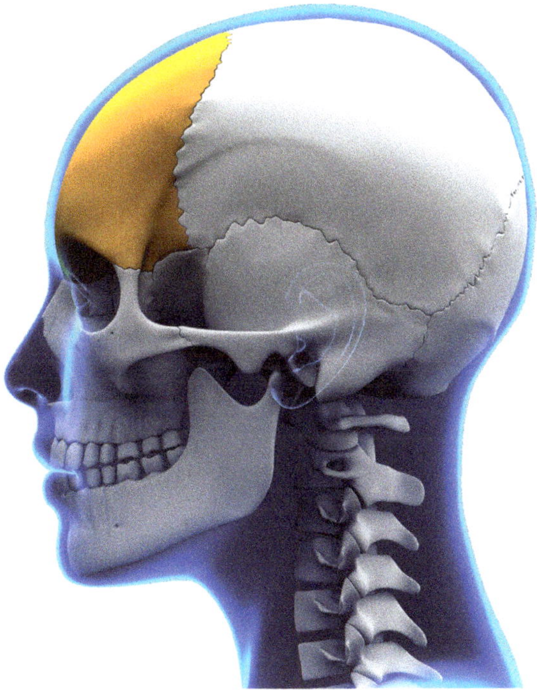

*Side View*

# Etymology

The word frontal comes from the Latin word *frons*, which means forehead.

# Bone Info

The frontal bone protects the brain, gives shape to the head, and supports the muscles of your face. This bone helps form the orbital and nasal cavities.

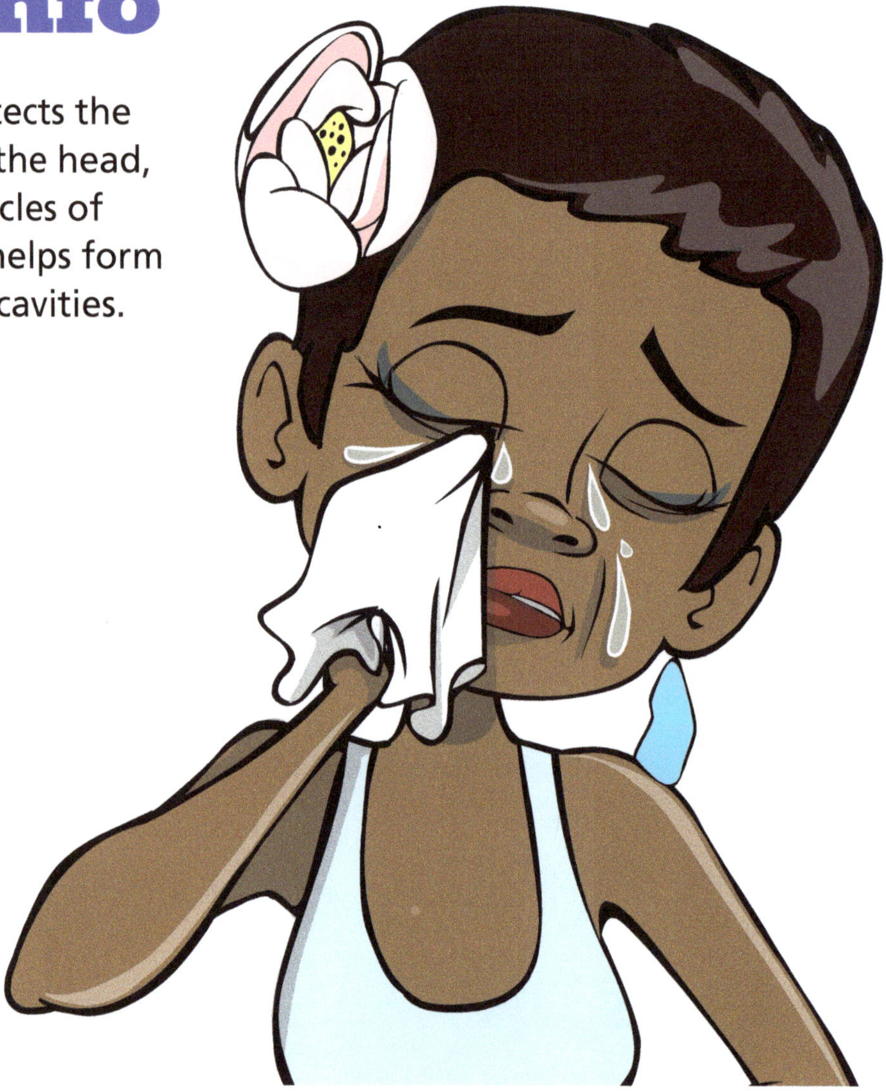

The specific part of the brain that the frontal bone protects is called the frontal lobe. The frontal lobe controls the muscles used for movement of the body. It also plays a role in problem solving, language, impulse control, and much more.

**Heads Up** *The glabella is a small, flattened portion of your frontal bone that joins the bony arches under your eyebrows. Can you feel where your glabella is?*

# Activity: Browse the Brows

The frontal bone doesn't move, but it's one of the most expressive areas of your face. The forehead has bands of muscle around it that contract and relax to move your eyebrows and the skin on your forehead.

Match the pictures of the characters below with the emotion that you think their brows and foreheads represent.

**Anger**

**Sadness**

**Happiness**

**Shock**

# Activity: Crossword: Bones of the Skull

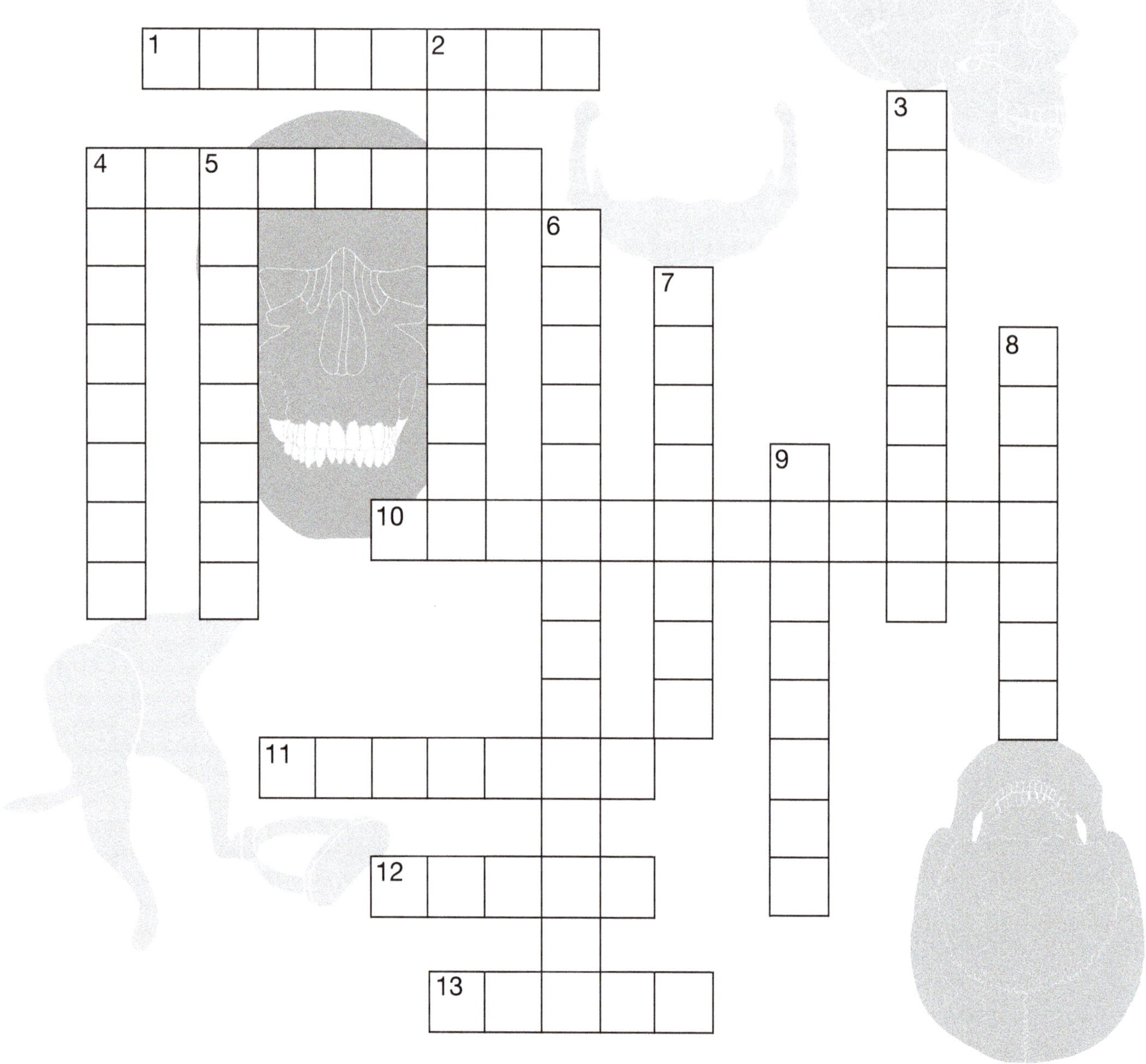


## across

1. Paired bone that gives the skull its dome shape
4. Humans have bilateral ______
10. Section of the skull that contains the eight cranial bones
11. The bowl-shaped bone that forms the forehead
12. The horseshoe-shaped bone in the neck that supports the tongue
13. The bone that looks like a plow

## down

2. Bone also known as nasal concha
3. Paired bone that makes up the most prominent part of the cheek
4. Bone that has "wings"
5. Another name for the jawbone
6. Section of the skull that contains the 14 facial bones
7. Term used to refer to all three bones of the ear
8. Paired bone that gets its name from the Greek word for "strainer"
9. Paired bone that gets its name from the Latin word for "tear"

# Activity: Pair It Up!

Remember that there are eight bones in the neurocranium, two pairs and four singles. Circle the paired bones, and leave the rest alone.

# Activity: Connect the Stars

Help Pinky connect the stars to form the shapes of two bones of the skull.

Can you tell which bones are in Skullapagos Island's sky?

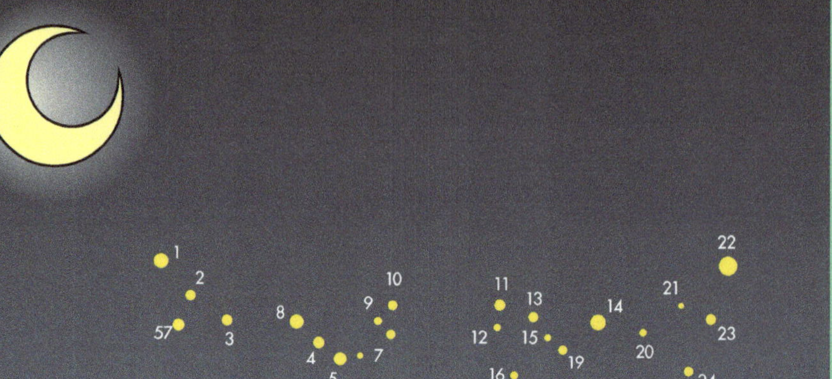

# Activity: Protect Your Head

Even though our skulls are built to protect our brains, when we ride bikes or skateboards, we always wear helmets like a second skull for extra protection. Which four bones of the neurocranium lie right under helmets?

1 ______________________

2 ______________________

3 ______________________

4 ______________________

Now, I just need to get to Stokely's party on time.
Good job remembering your helmet! Do you have an extra?

N
E
S
W

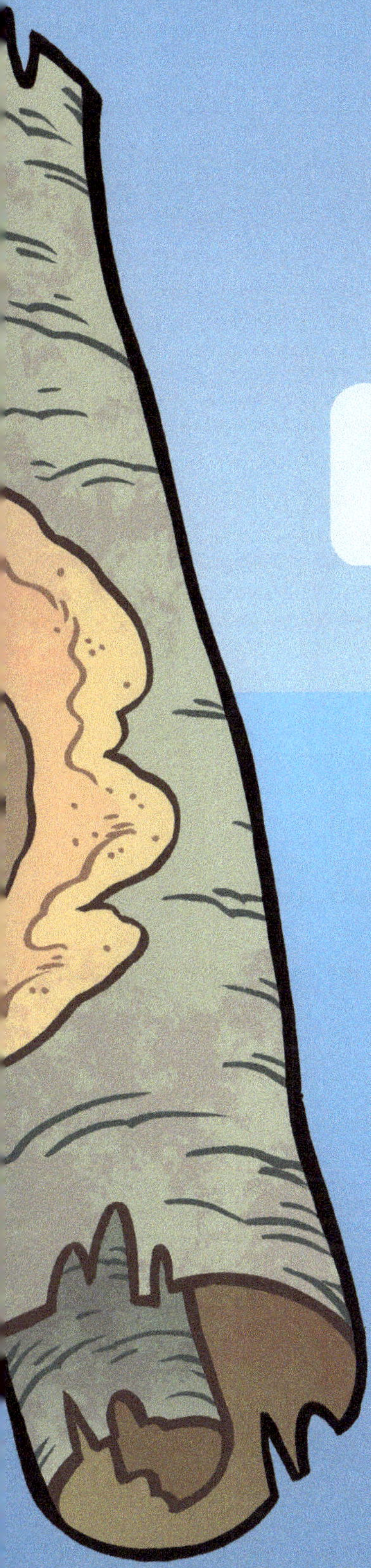

# Neurocranium Review

Use the mnemonic to help Pinky remember the bones of the neurocranium. Fill in the names of each bone in the spaces below.

S _ _ _ _ _ _ _

T _ _ _ _ _ _ _

E _ _ _ _ _ _

P _ _ _ _ _ _ _

O _ _ _ _ _ _ _ _

F _ _ _ _ _ _

# Activity: Colorful Calaveras

Decorative Mexican skulls, or *calaveras*, are traditionally made from sugar or clay. They are usually decorated with flowers, hearts, crosses, spiderwebs, paisleys, and leaves. Think about the shapes of the bones and which decoration will fit best.

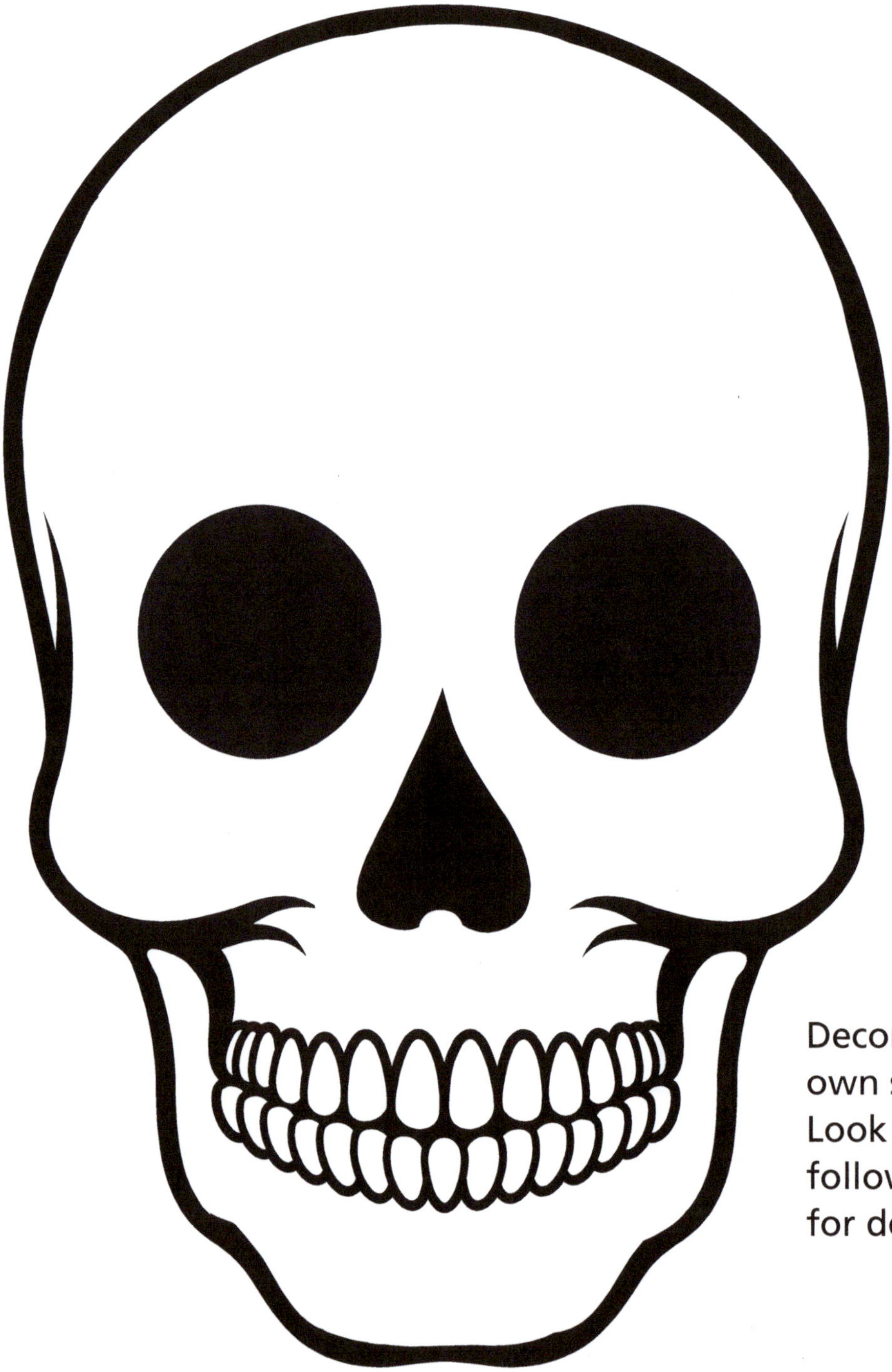

Decorate your own sugar skull. Look on the following page for design ideas.

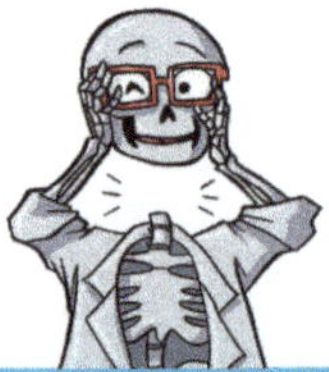

**Heads Up** *Día de los Muertos (Day of the Dead) is a Mexican holiday when people get together with friends and family to remember loved ones who have died. The sugar skull is a popular decoration.*

GREAT PARTY STOKELY!
Thanks, Pinky! It's perfect! It must have been really hard to make!
I just used my head!
FRONTAL

# We Did It!

*Well done, Bonyfide Buddies!*
*What dedication you've shown.*
*You're a head and face expert,*
*and your mind, it has grown.*

*You've learned 29 bones*
*of the structure inside.*
*Your knowledge has power.*
*You're now Bonyfide!*

*Wait! Before you head out,*
*there's more knowledge to gain.*
*It's some fun Bonus Knowledge*
*to tickle your brain.*

*You've earned one more surprise.*
*Check the back of this book.*
*It's your official certificate.*
*Go on, take a look!*

**Know Yourself Academy's**
*Bonus Knowledge*

Now that you've learned the bones of the head, face, and neck, step right up for some **Bonus Knowledge**! These pages will help you understand your body even more.

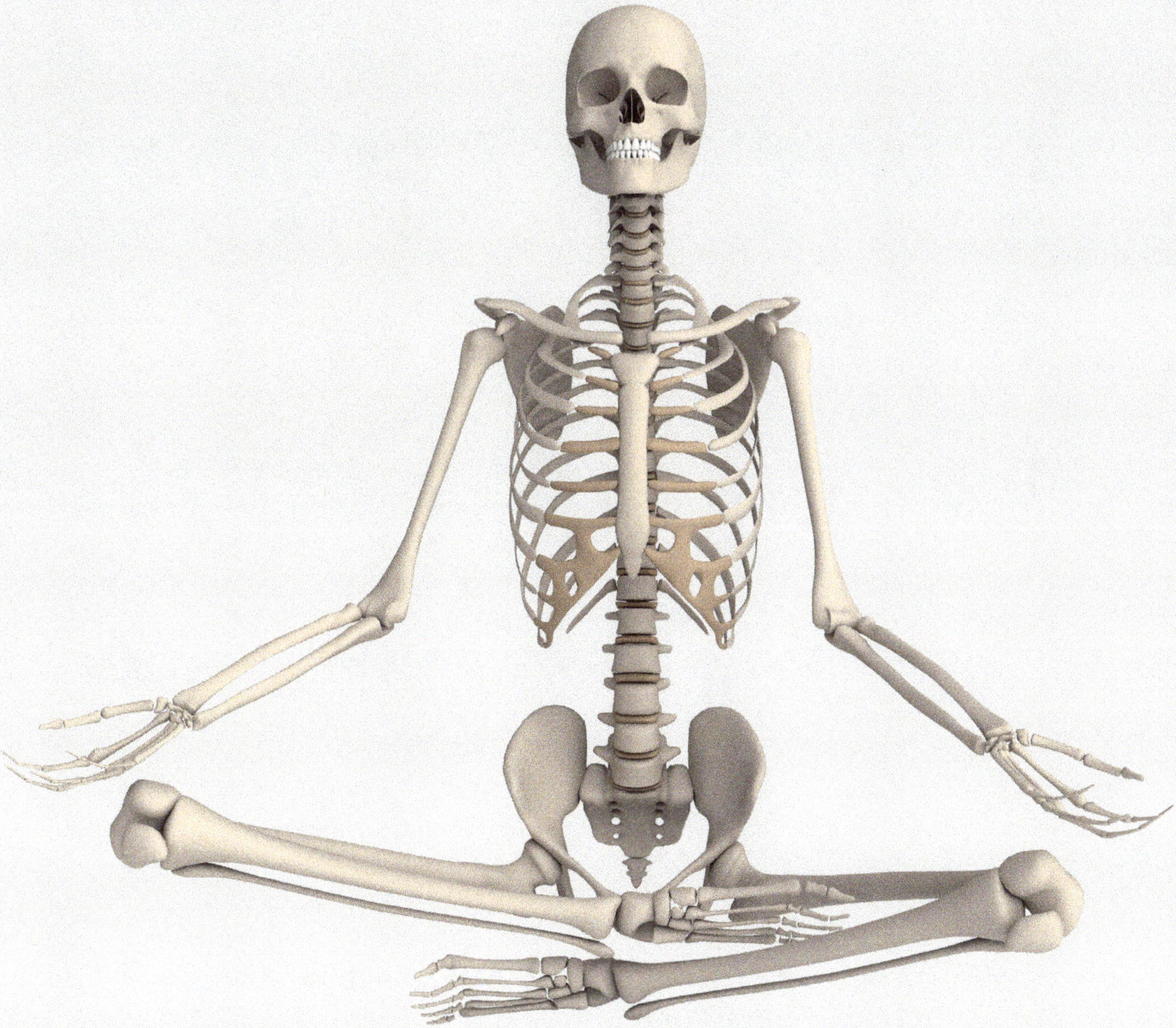

# The Lobes of the Brain

Several of the bones of the neurocranium are associated with a section of the brain, also called a lobe. Each lobe is a distinct area of the brain that controls certain essential human functions.

The parietal lobe is responsible for helping you process sensory input, such as taste, touch, pain, and temperature, so that you know where your body is and are aware of your surroundings. This can also be called spacial orientation, which is what allows you to navigate the world around you.

The frontal lobe is the boss of the brain. It's responsible for planning, focusing, solving problems, making decisions, and controlling impulses and motor functions. It's also considered the home of your personality, which continues to develop until you're an adult.

The occipital lobe is responsible for vision, helping you distinguish color and movement around you. It also helps with depth perception, allowing you to see patterns and how things are constructed.

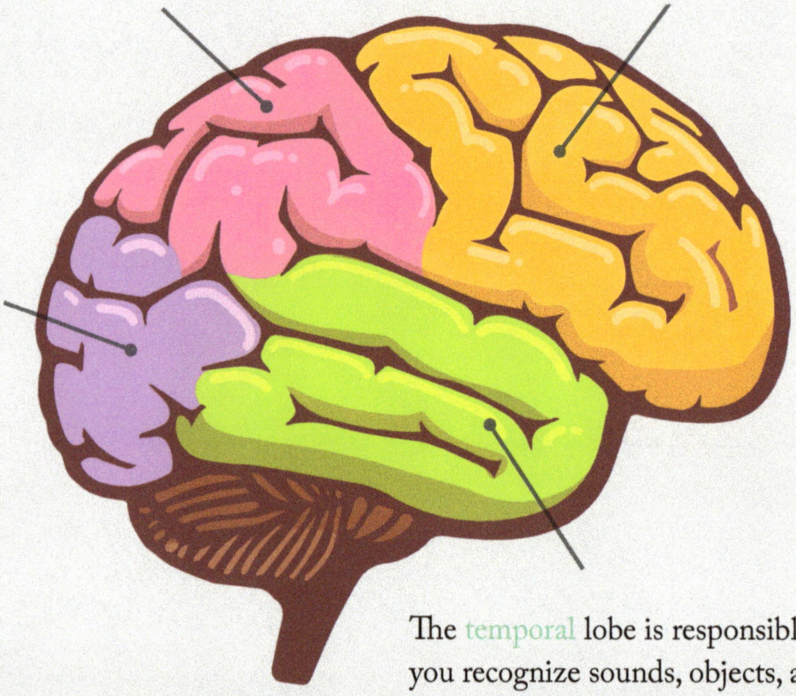

The temporal lobe is responsible for helping you recognize sounds, objects, and smells. It's also the language, emotional, and memory center of the brain, allowing you to name, feel, and remember important events in your life.

Underneath and in the back of your **cerebrum** (another word for all four of these lobes) is your cerebellum, which means "little brain." Although smaller than your cerebrum, your **cerebellum** contributes to balance, posture, muscle coordination, and motor learning.

Say it like this: "*suh-**ree**-brum*"
Say it like this: "*sehr-uh-**bell**-um*"

# Formation of the Skull

When humans are born, our skulls have many smaller pieces that slowly join into the bones we've covered in this book. Parents have to be very careful, because babies have several soft spots called **fontanelles** in the top and back of the head, areas where the skull bones haven't fused yet. After a baby is born, the posterior fontanelle stays open for about 2 to 4 months, to accommodate the baby's growing brain. The anterior fontanelle stays soft until the age of 18 months to 2 years.

As we get older, the bones fuse together (ossify) into one hard casing around the brain. Even after they join together, our skulls have fissures marking where the separate bones fused.

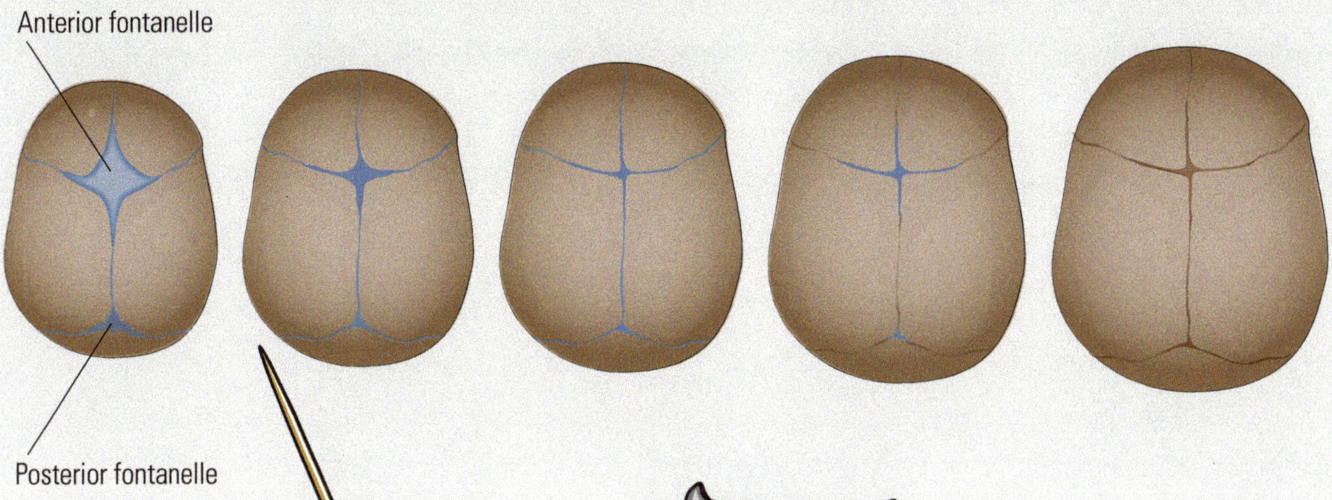

# Bone Groups

Sometimes, people refer to certain bones in the face by using a name that actually describes more than one bone. For example, have you ever heard of the orbital before? People use this term, or the phrase **orbital rim**, to discuss the eye socket. In fact, the orbital rim is comprised of seven different bones, from both the viscerocranium and neurocranium. They articulate, or connect together, to form this area. These seven bones are the maxilla, frontal, zygomatic, ethmoid, lacrimal, sphenoid, and palatine.

Some of these bones also help to form the **nasal cavity**, or the large air-filled space above and behind the nose. Specifically, parts of the maxilla, ethmoid, palatine, inferior nasal conchae (turbinates), nasal bones, vomer, and sphenoid are responsible for creating the nasal cavity's structure. Who knew breathing through your nose involved so many bones?!

# Are Teeth Bones?

Technically speaking, teeth are not bones. Like bones, they are made of calcium, phosphorus, and other minerals. However, teeth lack collagen, a living, growing tissue that gives bones their flexible framework and allows them to withstand pressure.

Babies are born with all the teeth they'll ever need. Most of us have 20 "baby teeth" and 32 permanent teeth, which show up after we lose our baby teeth in childhood.

In rare cases, some human babies are born with a third set of teeth. Sharks, on the other hand, have anywhere from 5 to 50 rows of teeth, and teeth are replaced thousands of times in their lives!

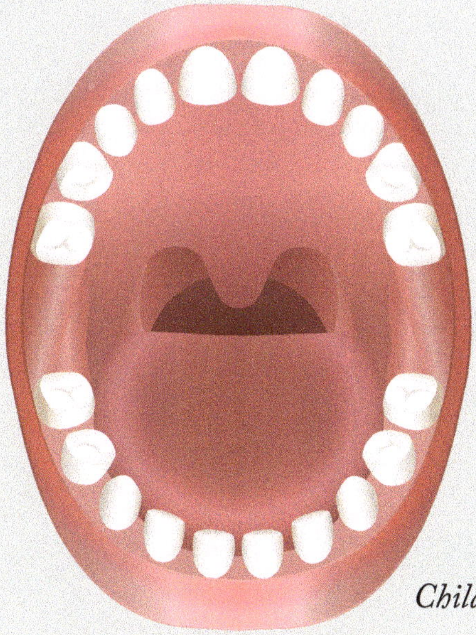

*Child*

*Adult*

# Turn it Down, Please!

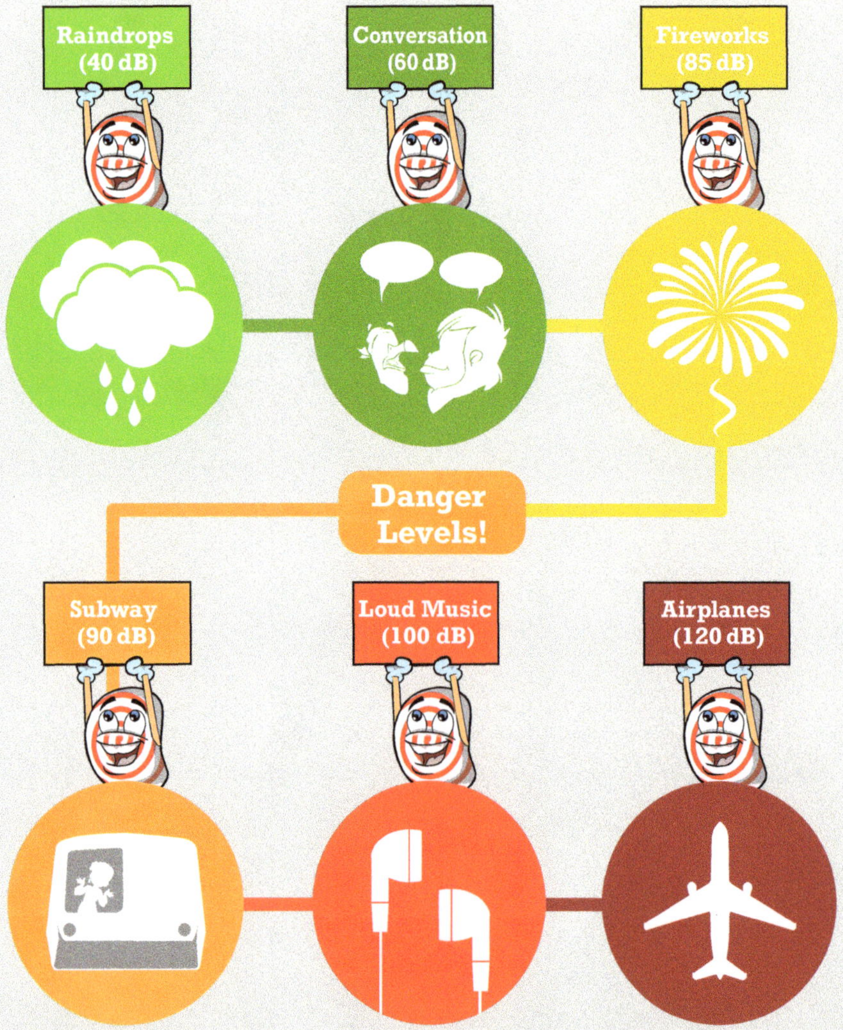


Sound is measured in units called decibels (dB). A **decibel** is a way of measuring the intensity of a sound wave.

The sounds an average person hears range from as low as 0 dB to over 180 dB (the sound of a rocket launch). Around 191 dB, sound waves become shock waves.

If you consistently expose yourself to decibel levels above 85 dB, you can damage your hearing. Sounds above 85 dB include sirens, concerts, jet engines, fireworks, chainsaws, car horns, and loud music in headphones. If you're going to be around loud noises like these, be sure to cover your ears, or get earplugs!

# Four Books, but How Many Bones?

Now that we've reached the end of our 4-book series on the skeletal system, it's time to count up all the bones we've learned about!

In **Book 1**, we learned about the bones of the hand, arm, and shoulder. In all, there were 64 bones.

In **Book 2**, we learned about the bones of the foot, leg, and pelvis. In all, there were 62 bones.

In **Book 3**, we learned about the bones of the rib cage and spine. In all, there were 51 bones.

In **Book 4**, we learned about the bones of the head, face, and neck. In all, there were 29 bones.

Add up each book's total to find out how many bones are in the whole human body!

$$\begin{array}{lr} & 64 \\ + & 62 \\ + & 51 \\ + & 29 \\ \hline = & \end{array}$$

# Glossary

**Bilateral Symmetry:** When something appears the same on two sides, like a mirror image. For example, bilateral symmetry is found throughout the human skull in the form of paired bones.

**Brain Cavity:** A large hollow space in the skull where the brain sits.

**Calavera:** A decorative Mexican skull, traditionally made from sugar or clay. Calaveras are usually decorated with flowers, squiggly lines, crosses, hearts, spiderwebs, paisleys, and leaves.

**Cartilage:** A strong but flexible connective tissue found in some parts of the body. It provides support without being as hard or rigid as bone. Most of your nose is made of cartilage.

**Cranium:** Another word for the neurocranium, or the section of the skull comprised of the bones that protect and house the brain and the delicate tissue around it.

**Dia de los Muertos (Day of the Dead):** A Mexican holiday in which people get together with friends and family to remember loved ones who have died. Calaveras are a popular decoration on Dia de los Muertos.

**Ethmoid:** The ethmoid is a square-shaped bone at the root of the nose, with many small openings for nerves to pass through. Ethmoid comes from the Greek word *ēthmoeidēs*, meaning "strainer." The air we breathe through the nose passes through the ethmoid.

**Etymology:** The study of the origin of words.

**Foramen Magnum:** The name of the large opening, found in the occipital bone, that the spinal cord passes through so that it can connect with the brain.

**Foramina:** Openings in certain bones that function as passageways for nerves and blood vessels.

**Frontal:** The frontal is a bowl-shaped bone that forms the front of the head. Frontal comes from the Latin word *frons*, which means forehead.

**Hyoid:** The hyoid is a horseshoe-shaped bone that supports the tongue and the larynx. It gets its name from the Greek word *huoiedēs*, which means "shaped like the letter *upsilon*," because it looks like a "U." Loosely held in place by a system of muscles and ligaments, it's the only bone in the human body that is not directly connected to any other bone. It's also the only neck bone considered part of the skull.

**Incus:** The incus is one of the three ear bones, collectively known as the ossicles. The shape of the incus resembles that of an anvil.

**Lacrimal:** Approximately the size of a fingernail, the lacrimal is a paired bone that sits inside the eye socket and plays an important role in the movement of tears. This is where the bone gets its name from.

**Larynx:** Also called the voicebox, the larynx is connected to the hyoid bone by several muscles. The structure of the larynx and the human hyoid bone makes speech possible.

**Malleus:** The malleus is one of the three ear bones, collectively known as the ossicles. The shape of the malleus resembles that of a hammer.

**Mandible:** The mandible is also known as the jawbone. It's the strongest, largest, lowest, and most moveable bone in the face. Mandible comes from the Latin word *mandere*, which means "to chew."

**Maxilla:** The maxilla is a paired bone that forms the upper jaw.

**Maxillae:** The plural form of maxilla.

**Microvilli:** Sensitive microscopic hair-like structures on the tongue that send messages to the brain, which the brain interprets as taste.

**Mnemonic:** A learning tool such as an acronym, rhyme, or song to help us remember information. "Mighty Pinky Moves Vertically Zip Lining Near Treasure" is an example.

**Nasal:** The nasal is a paired bone located in the upper middle part of the face. It helps form the "bridge" of the nose.

**Nasal Cavity:** A large air-filled space above and behind the nose in the middle of the face that is a continuation of the nostrils.

**Neurocranium:** The section of the skull comprised of the cranial bones. There are eight of these bones in all, two of them symmetrically paired, and four of them unpaired. The neurocranium protects and houses the brain and the delicate tissue around it.

**Occipital:** The occipital is a bone at the very back of the neurocranium. It forms part of the brain cavity, a large hollow space in the skull where the brain sits. It's the only bone of the skull that articulates with the cervical section of the spine.

**Ossicles:** The ossicles are the bones found in the ear (individually called the malleus, the incus, and the stapes).

**Osteologist:** A doctor who specializes in osteology.

**Osteology:** The branch of anatomy that deals with the structure and function of bones.

**Palatine:** The palatine is a paired bone located towards the back of the mouth. It helps to form part of the roof of the mouth and nasal cavity. Palatine comes from the Latin word *palatinus*, meaning "of the palace."

**Papillae:** Small bumps that protrude from the tongue. There are four types of papillae, three of which contain taste buds.

**Parietal:** The parietal is a paired bone that is the largest bone of the cranium and connects the front and back sections of the skull.

**Skeleton:** The bones that form the framework of the human body. Adult humans have 206 bones.

**Skull:** The bony structure that forms the head in the human skeleton. It supports the structure of the face and forms a space for the brain. The human skull protects the brain from injury.

**Sphenoid:** The sphenoid is a bone located in the middle of the skull. Its name comes from the Greek word *sphen*, or "wedge." Because of its shape, the outer parts of the sphenoid are called "wings." The sphenoid bone has openings called foramina that function as passageways for nerves and blood vessels.

**Spinal Cord:** The bundle of nerve fibers and tissue, housed by the spine, that carries messages between the brain and the rest of the body.

**Stapes:** The stapes is one of the three ear bones, collectively known as the ossicles. The shape of the stapes resembles that of stirrups.

**Temporal:** The temporal is a paired bone found behind and around the ears, helping to form the lower walls of the skull.

**Turbinate:** Another name for a nasal concha (plural, conchae). There are three kinds of nasal conchae: inferior, superior, and middle. The inferior nasal conchae are separate bones, but the superior nasal conchae and middle nasal conchae are part of the ethmoid bone.

**Viscerocranium:** The section of the skull comprised of the facial bones. There are 14 of these bones in all, six of them symmetrically paired and two of them unpaired.

**Vomer:** The vomer is the bone that separates the nasal passages, helping to direct the air you breathe. Vomer comes from the Latin word for "plowshare" because the bone has a similar shape to the traditional farming tool.

**Zygomatic:** The zygomatic is a paired bone that is found in the upper part of the face and makes up the most prominent part of the cheeks. Zygomatic comes from the Greek word *zygoma*, meaning "yoke."

# Answer Key

## Page 5

14 bones in the face + 8 bones in the head + 6 bones in both ears + 1 bone in the neck = 29 bones of the head, face, and neck

## Page 23

6 x 2 = 12; 1 + 1 = 2; 12 paired bones + 2 single bones = 14 viscerocranium bones

## Page 37

**Activity: Agree or Disagree?**
Seymour Seagull Wants to Caw-Know...

| Statement | Agree | Disagree |
|---|---|---|
| Seymour Seagull is Dr. Bonyfide's trusty sidekick. | ☑ | ☐ |
| The bones of the skull are classified in two main sections. | ☑ | ☐ |
| 14 + 8 + 6 + 1 = 30 | ☐ | ☑ |
| The mandible is a single bone in the viscerocranium. | ☑ | ☐ |
| A mnemonic is a memory tool that can be used to remember the names of the bones of the head, face, and neck. | ☑ | ☐ |
| The palatine is a famous palace. | ☐ | ☑ |
| People have bilateral symmetry, which means many bones of the skull are paired. | ☑ | ☐ |
| The Latin root for mandible refers to internal organs. | ☐ | ☑ |
| The maxilla is a paired bone that makes up the lower jaw. | ☐ | ☑ |

Mighty Pinky Moves Vertically Zip Lining Near Treasure

37

## Page 45

**Activity: Pair It Up!**
Remember that there are 14 bones in the viscerocranium, six pairs and two single bones.
So many bones below! Circle up the paired bones and leave the rest alone.


## Page 46

What other words can you make from the word palatine?
Some examples: nap, tale

What other words can you make from the word mandible?
Some examples: dim, name

## Page 47

WECH: chew
WGNA: gnaw
BBLEIN: nibble
CHNMU: munch
TIEB: bite
CPMOH: chomp

## Page 48-49

## Page 51

MANDIBLE
PALATINE
MAXILLA
VOMER

## Page 61

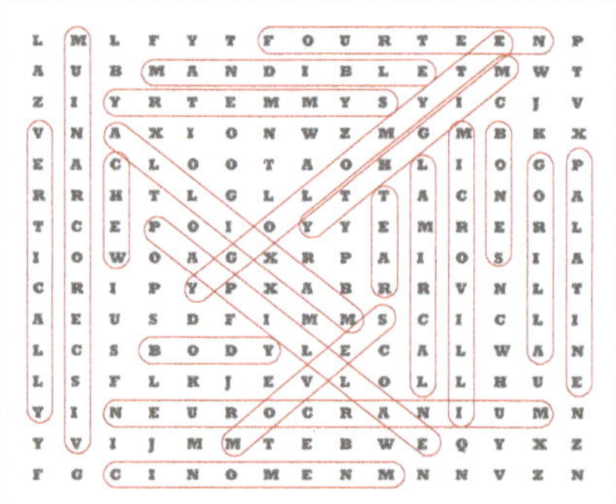

| | | | | | | | | | | | | | | |
|---|---|---|---|---|---|---|---|---|---|---|---|---|---|---|
| L | M | L | F | Y | T | F | O | U | R | T | E | E | N | P |
| A | U | B | M | A | N | D | I | B | L | E | T | M | W | T |
| Z | I | Y | R | T | E | M | M | Y | S | Y | I | C | J | V |
| V | N | A | X | I | O | N | W | Z | M | G | M | B | K | X |
| E | A | C | L | O | O | T | A | O | H | L | I | O | G | P |
| R | R | H | T | L | G | L | L | T | T | A | C | N | O | A |
| T | C | E | P | O | I | O | Y | Y | E | M | R | E | R | L |
| I | O | W | O | A | G | X | R | P | A | I | O | S | I | A |
| C | R | I | P | Y | P | X | A | B | R | R | V | N | L | T |
| A | E | U | S | D | F | I | M | M | S | C | I | C | L | I |
| L | C | S | B | O | D | Y | L | E | C | A | L | W | A | N |
| L | S | F | L | K | J | E | V | L | O | L | L | H | U | E |
| Y | I | N | E | U | R | O | C | R | A | N | I | U | M | N |
| Y | V | I | J | M | M | T | E | B | W | E | Q | Y | X | Z |
| F | G | C | I | N | O | M | E | N | M | N | N | V | Z | N |

## Page 75

The nasal conchae are also known as turbinates. The inferior nasal conchae are separate paired bones while the superior and middle nasal conchae are part of the ethmoid bone. The shape of nasal conchae make air slow down, or swirl when it enters the nose. This means air can be warmed and humidified, which adds moisture.

## Page 76

NASAL, ZYGOMATIC, MAXILLA, VOMER

## Page 77

The viscerocranium has a total of 14 bones.
What's another name for the jawbone? mandible
Which bone in the viscerocranium is shaped like a plow? vomer
What is the main function of this bone? to direct air into the nose
Which bone is also called the cheekbone? zygomatic
Why do you think we drew Lady Lacrima crying? What are we trying to help you remember? Sample answer: The lacrimal bone plays a role in the movement of tears. Drawing Lady Lacrima crying helps people remember this function.
What is another name for the nasal conchae? turbinates
What are two things we can learn about an animal by looking at its skull? Sample answer: what it eats and the animal's posture

## Page 79

ZYGOMATIC
LACRIMAL
NASAL
TURBINATE

## Page 88

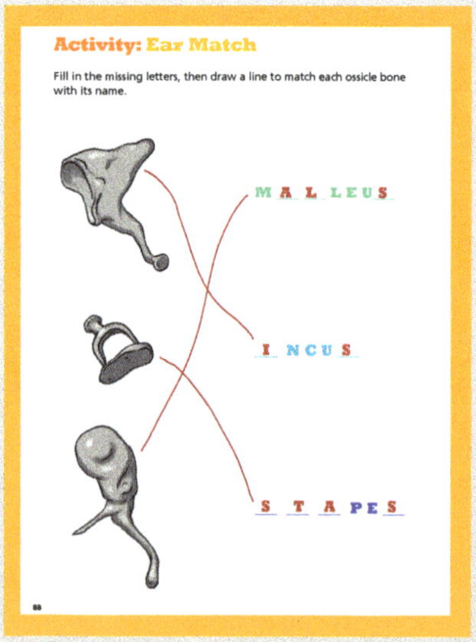
**Activity: Ear Match**
Fill in the missing letters, then draw a line to match each ossicle bone with its name.


88

## Page 93

Not connected to other bones
Helps humans talk
Only neck bone that's considered part of the skull

## Page 97

2 x 2 = 4; 4 x 1 = 4; 4 paired bones + 4 single bones = 8 neurocranium bones

## Page 106

| Names of the bones we know so far... | | 2 or 1 ? |
|---|---|---|
| M ANDIBLE | M | 1 |
| P ALATINE | P ALATINE | 2 |
| M AXILLA | M AXILLA | 2 |
| V OMER | V | 1 |
| Z YGOMATIC | Z YGOMATIC | 2 |
| L ACRIMAL | L ACRIMAL | 2 |
| N ASAL | N ASAL | 2 |
| T URBINATE | T URBINATE | 2 |

## Page 111

OS-SI-CLES
TUR-BI-NATE
NEU-RO-CRA-NI-UM
SPHE-NOID
MAL-LE-US
HY-OID
STA-PES

## Page 123

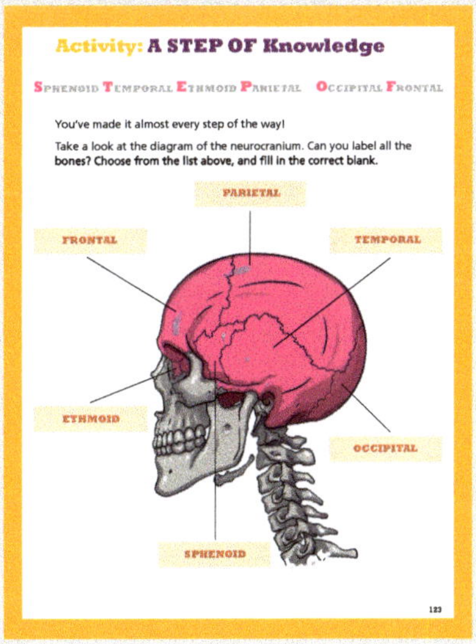
Activity: A STEP OF Knowledge
SPHENOID TEMPORAL ETHMOID PARIETAL OCCIPITAL FRONTAL
You've made it almost every step of the way!
Take a look at the diagram of the neurocranium. Can you label all the bones? Choose from the list above, and fill in the correct blank.
PARIETAL
FRONTAL
TEMPORAL
ETHMOID
OCCIPITAL
SPHENOID
123

## Page 130

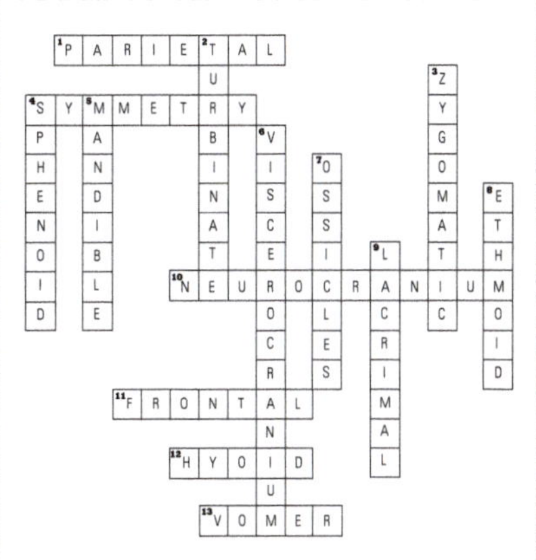


## Page 131

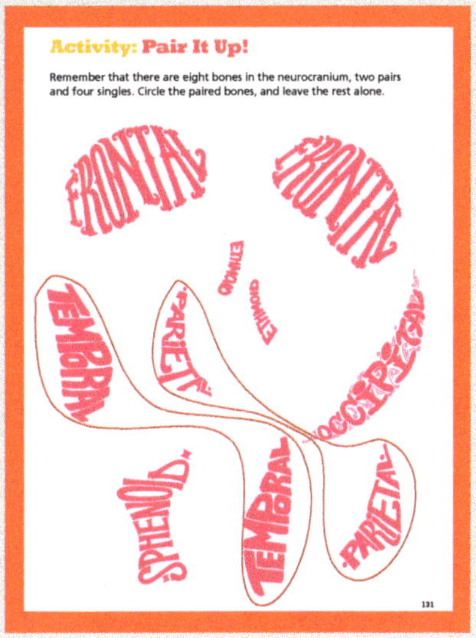
Activity: Pair It Up!
Remember that there are eight bones in the neurocranium, two pairs and four singles. Circle the paired bones, and leave the rest alone.
FRONTAL
FRONTAL
TEMPORAL
PARIETAL
OCCIPITAL
SPHENOID
TEMPORAL
PARIETAL
131

## Page 132

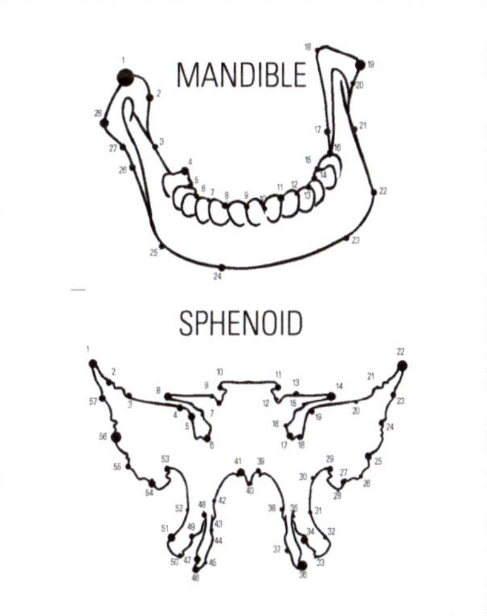
MANDIBLE
SPHENOID

## Page 133

PARIETAL, FRONTAL, OCCIPITAL, TEMPORAL

## Page 135

SPHENOID, TEMPORAL, ETHMOID, OCCIPITAL, FRONTAL

# Notes

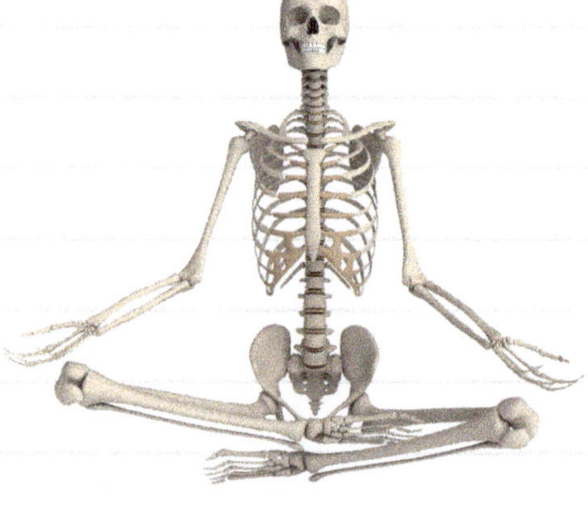

## Notes

# Notes

# Notes

# Notes

# Notes

# Notes

## Notes

The 3 Muska Tears
Oliver rejoined forces with Ladies Lacrima. **The 3 MuskaTears** are currently on tour.
Seymour and Occipital formed an unlikely friendship. They're planning on taking a cruise around the world this summer.

Maxilla used the plow Oliver made to give new life to her farm.
Pinky & Dr. B. continue their adventure in Book 5, "The Circulatory System."

# The Adventure Continues...

**If you like these books, check out our Anatomy Adventure Series to continue on your quest for Self Literacy!**

Self Literacy helps me understand how I move and balance, which makes me better at sports.
Understanding my mind and body helps me control my moods!
Knowing myself helps me understand my favorite ways to learn.
KNOW YOUR SELF
AND BE INSPIRED

**Self Literacy** [n.]: Having a working knowledge of your anatomy, physiology, and psychology.

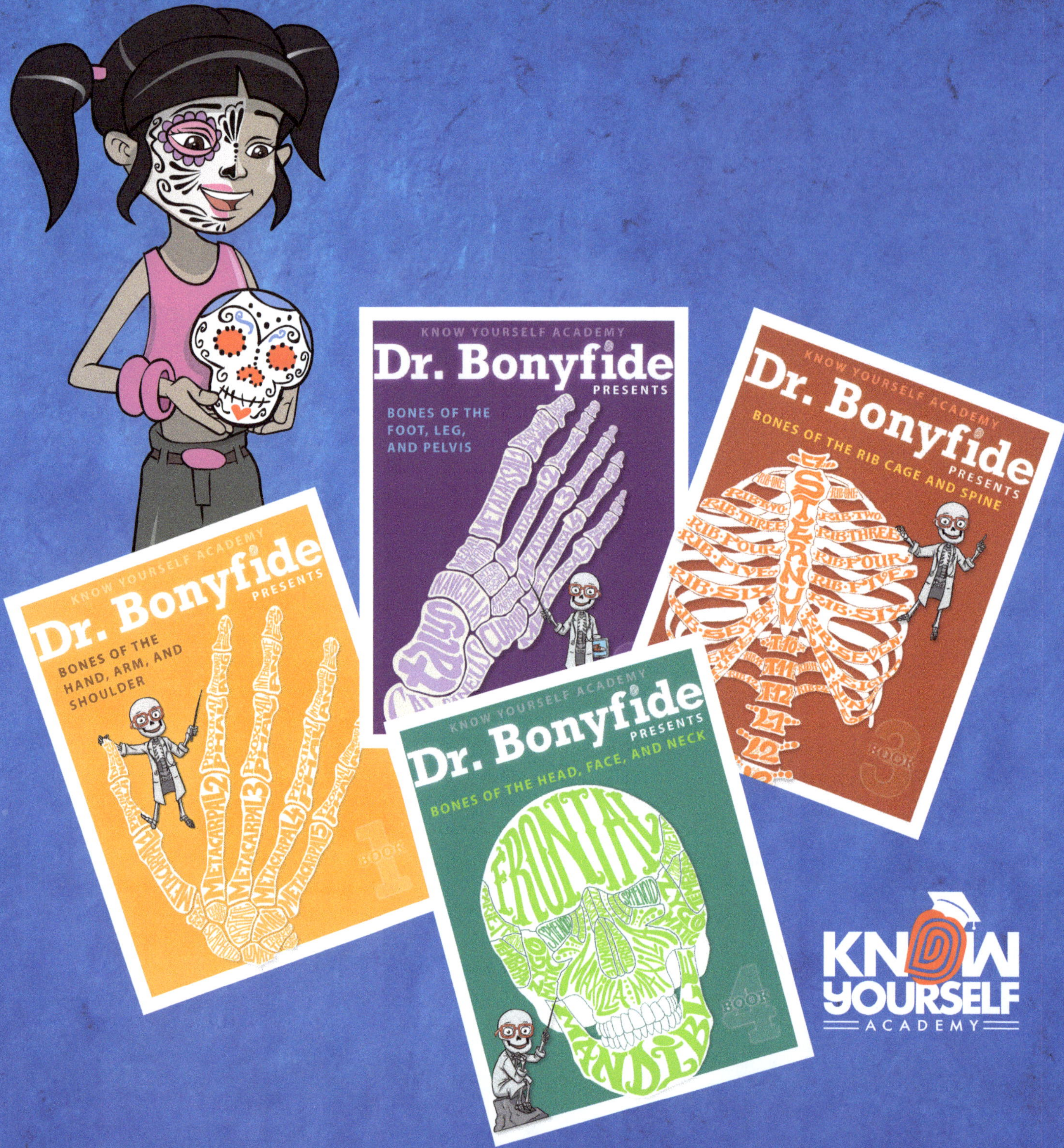


**Join us on the mission to Self Literacy and create the most awesome version of yourself.**

**Come find out at knowyourself.com**

www.ingramcontent.com/pod-product-compliance
Lightning Source LLC
Chambersburg PA
CBHW061410090426
42740CB00027B/3499
*9780991296835*